AF739776

An African Perspective on Environmental Communication

Given Mutinta

An African Perspective on Environmental Communication

Advocating for Sustainable Environmental Communication in Africa

palgrave
macmillan

Given Mutinta
Media and Communication Studies
Nelson Mandela University
Port Elizabeth, South Africa

ISBN 978-3-032-05051-9 ISBN 978-3-032-05052-6 (eBook)
https://doi.org/10.1007/978-3-032-05052-6

© The Editor(s) (if applicable) and The Author(s), under exclusive license to Springer Nature Switzerland AG 2025

This work is subject to copyright. All rights are solely and exclusively licensed by the Publisher, whether the whole or part of the material is concerned, specifically the rights of translation, reprinting, reuse of illustrations, recitation, broadcasting, reproduction on microfilms or in any other physical way, and transmission or information storage and retrieval, electronic adaptation, computer software, or by similar or dissimilar methodology now known or hereafter developed.
The use of general descriptive names, registered names, trademarks, service marks, etc. in this publication does not imply, even in the absence of a specific statement, that such names are exempt from the relevant protective laws and regulations and therefore free for general use.
The publisher, the authors and the editors are safe to assume that the advice and information in this book are believed to be true and accurate at the date of publication. Neither the publisher nor the authors or the editors give a warranty, expressed or implied, with respect to the material contained herein or for any errors or omissions that may have been made. The publisher remains neutral with regard to jurisdictional claims in published maps and institutional affiliations.

This Palgrave Macmillan imprint is published by the registered company Springer Nature Switzerland AG.
The registered company address is: Gewerbestrasse 11, 6330 Cham, Switzerland

If disposing of this product, please recycle the paper.

To my late mother, Georgina Silenga Mutinta, a woman for all seasons, who was the first to introduce me to environmental sustainability and communication through the way she led her life, despite her modest education.

Foreword

This book delves into the importance of communication in addressing environmental issues such as deforestation, pollution, climate change, and biodiversity loss. It highlights the media's role in shaping perceptions and understanding of environmental issues. In spite of this, however, the book also acknowledges the challenges faced by the media, such as limited resources, censorship, and lack of competence in environmental journalism. The book proposes that to overcome these limitations, traditional media such as radio, television, and newspapers; digital media such as Facebook, YouTube, X, and Instagram; and African traditional communication methods such as oral tradition and folklore could be used in environmental communication. The book posits that traditional media, digital media, and African traditional media have the potential to create platforms for grassroots activism, citizen journalism, and community involvement in tackling environmental issues. By employing these tactics, environmental communication initiatives and stakeholders can encourage positive behavioural changes. Involving diverse stakeholders, including government institutions, non-governmental organisations, businesses, and local communities, is crucial for creating effective and long-lasting solutions to environmental challenges. Collaboration and partnership are essential for fostering resilience and creativity in the face of environmental crises. Education is also vital in environmental communication, as integrating environmental considerations into school curricula and raising awareness among the populace can inspire the next generation of environmental communication specialists. The author urges readers to view this book as both an urgent call to sustainable environmental communication action and a roadmap for change, encouraging people from all walks of life to work together to achieve a more sustainable and equitable future for all living beings on the African continent, which is our home.

Harry Gwala District Municipality,
KwaZulu-Natal, South Africa

Ndabezitha Tenza

Preface

Communication is crucial in addressing environmental challenges, raising awareness, advocating for sustainable practices, and fostering community participation. However, African countries face challenges in disseminating information about environmental issues, leading to misconceptions and apathy. Media plays a significant role in bridging this gap but faces challenges such as limited resources. Traditional media outlets play important role in disseminating informative content, drawing people to critical environmental narratives.

Digital media has emerged as a potent tool for environmental communication, allowing for rapid dissemination of information and dialogue among stakeholders. Grassroots campaigns often use digital platforms like X and Facebook to mobilise support for environmental initiatives. However, traditional African communication methods, such as oral tradition and folklore, have the potential to promote culturally relevant environmental messages. Integrating African traditional communication practices with modern strategies can enhance the effectiveness of environmental communication initiatives.

Stakeholder collaboration in environmental communication between governments, non-governmental organisations, and local communities can foster a more inclusive approach to addressing environmental communication challenges. Partnerships can enhance resource sharing and facilitate educational initiatives that empower communities to take communication action.

Education is a cornerstone of sustainable environmental communication and incorporating it into school curricula and community programmes can foster a culture of sustainability.

Besides, monitoring and evaluation are essential components of environmental communication programmes. They not only allow organisations to track progress and assess impact but also promote accountability and foster continuous improvement. As society faces increasingly complex environmental challenges, effective monitoring and evaluation will be vital in ensuring that communication strategies lead to meaningful change.

To address environmental communication challenges Africa is facing, policies must prioritise the integration of sustainable communication strategies, enhance media literacy, and support both formal and informal educational initiatives.

Port Elizabeth, South Africa Given Mutinta

Acknowledgements

I would like to begin by acknowledging my sons. Balancing family life with academic or creative pursuits can often be challenging. However, my sons have made this journey enjoyable and fulfilling. Their encouragement during holiday time has been invaluable. For instance, during the writing of this project, they would often ask about my progress. Their simple yet profound questions challenged me to think critically and reinforced my commitment to my work. This familial support allowed me to navigate through successes and setbacks with a positive outlook.

Furthermore, the role of friends in my life cannot be overstated. Their diverse perspectives and honest feedback have helped to refine my ideas. During discussions over coffee or casual meetups, my friends would share their views, pushing me to explore avenues I had not considered. Their collective input not only enriched my understanding but also boosted my poise, leading me to produce a more polished final piece.

In addition to my immediate support network, I must also recognise the contributions of unknown reviewers. These reviewers provided an objective perspective, pointing out areas for improvement that I may have overlooked. Their insights compelled me to refine my arguments and structure, ensuring that my work meets the highest standards of clarity and coherence. This external critique was invaluable, as it exposed my work to a broader audience and enhanced its credibility.

Ethics Approval Ethical considerations were maintained since this study utilised secondary data. This was essential to guarantee that the author adheres to the principles of honesty, respect for individuals, and robust ethical conduct. By conducting a thorough evaluation of the origins and contexts of secondary data, and by adhering to ethical standards, the author obtained permission to utilise particular material and properly acknowledged all sources employed. This was implemented to augment the body of knowledge in environmental communication while maintaining audience trust. Adhering to ethical standards in this book enhanced its reputation and promoted responsible scholarship.

Competing Interests The author has no competing interests to declare that are relevant to the content of this manuscript.

About This Book

This book explores the role of the media in environmental communication within the African context. It sheds light on the various challenges and opportunities that African countries face as they strive to address pressing environmental concerns. It highlights how the effects of climate change, deforestation, and pollution have heightened the need for effective communication strategies to raise awareness and foster sustainable practices among communities.

The book explores the critical importance of sustainable environmental communication in addressing environmental issues. It argues that effective communication can educate communities about environmental challenges, mobilise them for action, and ultimately transform behaviour towards more sustainable practices. The role of media—both traditional and digital media—is emphasised. This book highlights various successful campaigns that have utilised different media platforms to raise awareness about environmental issues, illustrating the potential of media to foster community participation and activism. It also highlights some Africans who have contributed to the struggle for environmental sustainability.

Furthermore, the book investigates traditional African communication methods, emphasising their relevance in contemporary environmental messaging. These methods, characterised by storytelling, dances, songs, proverbs, and community gatherings, can effectively convey complex environmental issues in ways that resonate with local cultures. Incorporating these traditional practices into modern communication strategies can enhance the effectiveness of environmental messages.

The book also underscores the importance of stakeholder involvement in environmental communication, outlining how collaboration among governments, NGOs, community leaders, and academia can create a more unified approach to tackling environmental challenges. By building partnerships, stakeholders can share resources, expertise, and networks, fostering a more robust and effective communication response to environmental issues.

Education is another critical component discussed in the book. It highlights the need for environmental education at all levels to empower individuals and communities with the knowledge and skills necessary to contribute to sustainability efforts.

Besides, the book highlights the importance of monitoring and evaluation in environmental communication programmes, serving as essential tools to measure effectiveness and impact. The book concludes with actionable policy recommendations, advocating for integrating sustainable environmental communication into national policies, investing in media capacity building, and enhancing educational curricula to address environmental issues comprehensively.

Contents

1 **Introduction to Environmental Communication** 1
1.1 The Concept of Environmental Communication 1
1.2 History of Environmental Communication 3
1.2.1 Factors That Led to the Emergency of Environmental Communication 4
1.3 Principles of Effective Environmental Communication 8
1.3.1 Understanding the Audience 9
1.3.2 Use Clear and Simple Language 10
1.3.3 Credible Sources 10
1.3.4 Making Use of Multiple Channels 11
1.3.5 Promote Bidirectional Dialogue 12
1.3.6 Tailored Messaging 13
1.3.7 Using Visual Aids 13
1.3.8 Culturally Sensitivity Communication 14
1.3.9 Empowering Messages 15
1.3.10 Repetitive and Consistency Messaging 15
1.3.11 Evaluating and Adapting 16
1.3.12 Interactive Communication 17
1.3.13 Sustainable Communication 17
1.3.14 Entertainment Education 19
1.3.15 Social Change 20
1.4 The Importance of Environmental Communication 21
1.4.1 Empowering Marginalised Communities 21
1.4.2 Enhance Sustainability Behaviour 22
1.4.3 Promoting Partnerships 22
1.4.4 Bridging the Gap 23
1.4.5 Foster Public Participation 24
1.4.6 Encourage the Sharing of Knowledge 24

1.4.7 *Encourage Capacity-Building* 25
1.4.8 *Increasing Awareness* 26
1.4.9 *Conclusion* 26
References 27

2 Environmental Challenges in Africa 29
2.1 *The Concept of Environment* 29
2.2 *Environmental Challenges* 31
2.2.1 *Deforestation* 31
2.2.2 *Land Degradation* 31
2.2.3 *Urbanisation* 33
2.2.4 *Water Scarcity* 34
2.2.5 *Water Pollution* 35
2.2.6 *Climate Change* 36
2.2.7 *Biodiversity Loss* 37
2.2.8 *Habitat Destruction* 38
2.2.9 *Air Pollution* 39
2.2.10 *Natural Disasters* 40
2.2.11 *Energy Accessibility* 41
2.2.12 *Desertification* 42
2.2.13 *Wildlife Trafficking* 43
2.2.14 *Plastic Pollution* 44
2.2.15 *Energy Insecurity* 45
2.3 *Conclusion* 46
References 47

3 The Role of the Media in Environmental Communication 51
3.1 *The Role of the Traditional Media in Environmental Communication* 51
3.1.1 *Television* 52
3.1.2 *Radio* 53
3.1.3 *Newspaper* 57
3.1.4 *Magazine* 59
3.1.5 *Posters* 60
3.1.6 *Billboards* 60
3.1.7 *Newsletters* 61
3.1.8 *Journal Articles* 62
3.1.9 *Books* 63
3.2 *Digital Platforms* 64
3.2.1 *Social Media* 64
3.3 *Successful Campaigns for Environmental Communication* 73
3.3.1 *Trashy Bags* 73
3.3.2 *The Green Belt Movement* 74
3.3.3 *The Plastic Agenda Campaign* 76

3.3.4 *Clean Up Kenya* 77
3.3.5 *The Green Legacy Initiative* 78
3.3.6 *The Eco Warriors of Ghana* 80
3.3.7 *Save the Seas of Seychelles* 81
3.3.8 *Greening Egypt* 82
3.4 *Indigenous Individual Environmental Advocates* 83
3.4.1 *Ken Saro-Wiwa* 83
3.4.2 *Wangari Maathai* 84
3.4.3 *Vanessa Nakate* 86
3.4.4 *Desmond D'Sa* 87
3.5 *Conclusion* 89
References 90

4 Role of the Traditional African Communication Methods in Environmental Communication 93
4.1 *Oral Traditions* 93
4.1.1 *Storytelling* 94
4.1.2 *Legends* 94
4.1.3 *Myths* 96
4.1.4 *Folk Tales* 97
4.2 *Folklore* 98
4.2.1 *Proverbs* 98
4.2.2 *Riddles* 99
4.2.3 *Chants* 100
4.2.4 *Songs* 101
4.2.5 *Poetry* 102
4.2.6 *Oral History* 102
4.2.7 *Proclamations and Announcements* 103
4.2.8 *Tribal Meetings* 104
4.2.9 *Elders and Wisdom* 105
4.2.10 *Rituals* 107
4.2.11 *Dances* 108
4.3 *Conclusion* 109
References 110

5 Challenges and Opportunities for Environmental Communication 113
5.1 *Challenges to Environmental Communication* 113
5.1.1 *Lack of Resources* 113
5.1.2 *Lack of Infrastructure* 114
5.1.3 *Language Barriers* 115
5.1.4 *Limited Access to Information* 116
5.1.5 *Government Censorship and Media Restrictions* 116
5.1.6 *Lack of Knowledge* 117

5.1.7 *Cultural Norms and Values* 118
5.1.8 *Denialism About Climate Change* 119
5.1.9 *Lack of Collaboration* 119
5.1.10 *Lack of Motivation* 120
5.1.11 *Cultural Differences and Beliefs That Influence Environmental Communication* 121
5.1.12 *Misinformation and Disinformation* 121
5.1.13 *Lack of Governmental Commitment and Will* 122
5.2 *Opportunities for Environmental Communication in Addressing Environmental Issues* 123
5.3 *Conclusion* 126
References 127

6 Stakeholder Engagement in Environmental Communication 129
6.1 *The Concept of Stakeholder Engagement* 129
6.2 *The Importance of Stakeholder Engagement in Environmental Communication* 131
6.3 *Stakeholders* 132
6.3.1 *Non-governmental Organisations* 132
6.3.2 *Governmental Agencies* 133
6.3.3 *Civil Society Organisations* 133
6.3.4 *Researchers* 134
6.3.5 *Community Leaders* 135
6.3.6 *Faith-Based Organisations* 136
6.3.7 *Media Organisations* 136
6.3.8 *Individuals at the Grassroots Level* 137
6.4 *Conflicts Among Environmental Advocates* 138
6.4.1 *Industry Versus Environmentalists* 138
6.4.2 *Government Versus Citizens* 139
6.4.3 *Local Communities Versus Conservationists* 140
6.4.4 *Consumers Versus Producers* 141
6.4.5 *Scientists Versus Sceptics* 142
6.4.6 *Urban Versus Rural* 143
6.5 *Conclusion* 144
References 145

7 The Role of Environmental Communication in Education 149
7.1 *The Importance of Environmental Communication in Education* 149
7.1.1 *Advocacy and Activism* 150
7.1.2 *Public Relations and Marketing* 151
7.1.3 *Community Engagement* 152
7.1.4 *Political Advocacy* 152
7.2 *Storytelling* 153
7.3 *Collaboration* 154

7.4 *Building Resilience to Climate Change* 155
7.5 *Promoting Environmental Justice* 155
7.6 *Mobilising Support for Environmental Initiatives* 156
7.7 *Integrating Environmental Communication into Educational Curricula* 157
7.7.1 *Raising Awareness* 157
7.7.2 *Promoting a Sense of Responsibility* 158
7.7.3 *Encouraging Critical Thinking* 158
7.7.4 *Promoting Advocacy* 159
7.7.5 *Building Communication Skills* 160
7.7.6 *Promoting Creativity* 160
7.7.7 *Connecting Classroom Learning to Environmental Issues in the Real World* 161
7.7.8 *Promoting Interdisciplinary Learning* 162
7.7.9 *Promoting Collaboration* 162
7.7.10 *Cultivating Knowledge* 163
7.8 *Case Studies of Initiatives for the Integration of Environmental Issues in Education* 164
7.8.1 *The Eco-School Programme* 164
7.8.2 *Green Clubs* 164
7.8.3 *Outdoor Education* 165
7.8.4 *The Wildlife and Environment Society of South Africa* 166
7.9 *Conclusion* 167
References 168

8 Monitoring and Evaluating the Impact of Environmental Communication Strategies 171
8.1 *The Importance of Monitoring and Evaluating Environmental Communication Programmes* 171
8.1.1 *Impact on the Target Audience* 172
8.1.2 *Identifying Potential Barriers* 172
8.1.3 *Maximising Resources and Cost-Effectiveness* 173
8.1.4 *Building a Strong Evidence Base* 174
8.1.5 *Ascertaining the Impact of the Messages and Communication Strategies* 175
8.1.6 *Promoting Accountability and Transparency* 176
8.1.7 *Opportunities of Innovation* 176
8.1.8 *Sustaining Long-Term Impacts* 177
8.2 *Theoretical Monitoring and Evaluation* 178
8.2.1 *The Health Belief Model* 178
8.2.2 *The Stages of Change Model* 179
8.2.3 *Knowledge-Deficit Model* 181
8.2.4 *Social Marketing Model* 181
8.2.5 *The Diffusion of Innovation Theory* 182

8.2.6 *The Framing Theory* 183
8.2.7 *The Agenda-Setting Theory* 184
8.2.8 *The Participatory Communication Model* 185
8.3 *Conceptual Metrics for Assessing the Impact of Environmental Communication Programmes* 186
8.3.1 *Awareness* 186
8.3.2 *Knowledge* 187
8.3.3 *Attitudes* 188
8.3.4 *Behaviour Change* 189
8.3.5 *Engagement* 189
8.3.6 *Reach* 190
8.3.7 *Impact on Environmental Outcomes* 190
8.4 *Conclusion* 191
References 192

9 Conclusions, Policy, Recommendations, and Future Direction to Overcome Environmental Communication Challenges 197
9.1 *Conclusions* 197
9.1.1 *Concept of Environmental Communication* 197
9.1.2 *Environmental Challenges* 198
9.1.3 *The Role of Traditional and Digital Media in Environmental Communication* 198
9.1.4 *Traditional African Communication Tools* 199
9.1.5 *Challenges and Opportunities for Environmental Communication* 199
9.1.6 *The Role of Stakeholders in Environmental Communication* 200
9.1.7 *The Role of Environmental Communication in Education* 200
9.2 *The Importance of Policy in Environmental Communication* 201
9.2.1 *Tailoring Messaging to Local Realities* 202
9.2.2 *Using the Traditional Channels of Communication* 203
9.2.3 *Emphasising the Benefits of Sustainability* 203
9.2.4 *Promoting Two-Way Communication* 204
9.2.5 *Utilising Digital Technologies* 205
9.2.6 *Building Capacity and Skills* 205
9.2.7 *Promoting Storytelling and Visuals* 206
9.2.8 *Promoting Collaboration and Networking* 207
9.2.9 *Monitoring and Evaluating Impacts* 207
9.3 *Recommendations* 208
9.3.1 *Government Agencies* 208
9.3.2 *Policymakers* 209
9.3.3 *Media Organisations* 210
9.3.4 *Civil Society Organisations* 211

9.3.5 Researchers 212
9.3.6 Educators 213
9.3.7 Community Leaders 213
9.3.8 Sponsors of Environmental Communication 214
9.3.9 Curriculum Developers for Educational Institutions 215
9.4 Future Direction 216
9.5 Conclusion 217
References 218

Glossary 221

Index 223

About the Author

Given Mutinta is an associate professor in the Department of Media and Communication Studies at Nelson Mandela University, South Africa. He holds two doctorates: one in Media and Communication Studies and the other in Research Methodology, both obtained from the University of KwaZulu-Natal. He teaches several courses in media and communication studies, and in research methodology. His research interests include health communication, communication for behavioural and social change, corporate communication, communication for development, strategic communication, public relations, and research methodology. His specific research focuses on the use of action research to advance communication-centred approaches in the awareness of societal issues and integrated solutions.

1st Email: for publication purposes: Gmutinta@mandela.ac.za

2nd Email: for correspondence purposes: Gmutinta@mandela.ac.za

List of Figures

Fig. 1.1 Industrialisation. Courtesy of Wikimedia Commons (2025) 5
Fig. 1.2 Globalisation. Artificial Intelligence Generated (2025) 6
Fig. 1.3 Natural disaster. Courtesy of Wikimedia Commons (2025) 8
Fig. 1.4 Bidirectional dialogue. Author (2025) 12
Fig. 1.5 Interactive communication. Courtesy of Freepik (2025) 18
Fig. 1.6 Entertainment education. Courtesy of Wikimedia Commons (2025) 20
Fig. 2.1 Land degradation. Courtesy of Wikimedia Commons (2025) 32
Fig. 2.2 Urbanisation. Courtesy of Freepik (2025) 33
Fig. 2.3 Water scarcity. Courtesy of Wikimedia Commons (2025) 34
Fig. 2.4 Water pollution. Courtesy of SDCEA (2024) 36
Fig. 2.5 Climate change. Courtesy of Wikimedia Commons (2025) 37
Fig. 2.6 Air pollution. Courtesy of Freepik (2025) 40
Fig. 2.7 Desertification. Courtesy of Wikimedia Commons (2025) 42
Fig. 2.8 Wildlife trafficking. Courtesy of Wikimedia Commons (2025) 43
Fig. 2.9 Plastic pollution. Courtesy of SDCEA (2024) 45
Fig. 3.1 Radio image. Artificial Intelligence Generated (2025) 54
Fig. 3.2 Newspapers. Courtesy of Wikimedia Commons (2025) 58
Fig. 3.3 Billboard. Courtesy of Wikimedia Commons (2025) 61
Fig. 3.4 Books. Courtesy of Freepik (2025) 64
Fig. 3.5 Image of Ken Saro-Wiwa. Courtesy of Noo Saro-Wiwa (2025) 84
Fig. 3.6 Image of Wangari Maathai. Courtesy of Wanjira Mathai (2025) 85
Fig. 3.7 Image of Vanessa Nakate. Courtesy of Nakate and Credit of Laura Cook (2025) 87
Fig. 4.1 Storytelling. Artificial Intelligence Generated (2025) 95
Fig. 4.2 Folk tale imaginary image. Courtesy of Wikimedia Common (2024) 98
Fig. 4.3 Singing. SDCEA (2024) [24] 101
Fig. 4.4 Tribal meetings. Courtesy of Wikimedia Common (2022) 105
Fig. 4.5 Elders and wisdom. Artificial Intelligence Generated (2024) 106
Fig. 4.6 Rituals. Courtesy of Wikimedia Common (2022) 107
Fig. 4.7 Dance. SDCEA (2024) [24] 109

Fig. 8.1 The health belief model. Author (2025) 179
Fig. 8.2 Stages of change model. Author (2025) 180
Fig. 8.3 The diffusion of innovation theory. Author (2025) 183
Fig. 8.4 The agenda-setting theory. Author (2025) 184
Fig. 8.5 Conceptual metrics. Author (2025) 186

CHAPTER 1

Introduction to Environmental Communication

Introduction
Environmental communication is the process of informing various audiences about environmental issues, challenges, and solutions. With time, environmental communication has evolved and is now crucial for changing attitudes, influencing behaviours, and raising public awareness of environmental preservation and sustainability. This chapter introduces the concept of environmental communication. The first part of the chapter covers the concept and history of environmental communication. The second part examines the factors that influenced the development of environmental communication as well as the effective principles of environmental communication. By understanding what constitutes environmental communication, organisations can effectively communicate about environmental issues and encourage action at the individual, interpersonal, community, and societal levels.

1.1 The Concept of Environmental Communication

The word "environment" originates from the French word "environ," which means "to encircle" or "to surround." This word reflects the fundamental concept that the environment encompasses everything around us, shaping our lives and experiences. The root of "environ" can be traced back to the Latin term "invirare," which means "to encircle" or "surround." This Latin word is derived from "virare," meaning "to turn," indicating a movement or action that creates a boundary [1]. Therefore, the evolution of the understanding of "environment" encapsulates the idea of a surrounding space, emphasising the interconnectedness of various elements within our surroundings [2]. On the

© The Author(s), under exclusive license to Springer Nature Switzerland AG 2025
G. Mutinta, *An African Perspective on Environmental Communication*,
https://doi.org/10.1007/978-3-032-05052-6_1

other hand, the word "communication" comes from the Latin word "communicare," which means to share or make common [1]. Environmental communication, then, is the dissemination of knowledge, information, ideas, and messages about the environment and related concerns. Initially, environmental communication was a restricted domain; however, it has since developed into an extensive field that includes research and practices related to the interactions among various entities, such as institutions, governments, and individuals, regarding environmental issues, as well as the influence of cultural products on societal attitudes towards these matters [1]. Thus, environmental communication is complex, including verbal, nonverbal, personal, interpersonal, and visual methods. Environmental communication involves human interactions with the environment. This includes several possible interactions, such as human engagement, virtual communities, participatory decision-making, and environmental media reporting. Some scholars define environmental communication as the use of communication approaches, ideas, strategies, and tactics within the context of environmental management and conservation [1]. Researchers contend that an individual's perspective on the world shapes their beliefs in many ways, facilitating the understanding of how environmental communication affects individuals. The notion that nature "communicates" is essential to the discipline of environmental communication. Various theories exist in this domain to elucidate the principles of environmental communication [2].

Environmental communication, as an academic field, emerged from interdisciplinary cooperation among communication, environmental studies, environmental science, risk analysis and management, sociology, and political ecology. In his 2004 textbook, Alexander Flor considers environmental communication an essential element of the environmental sciences, which he views as transdisciplinary. The author starts his textbook on environmental communication with a clear declaration: "Environmentalism, as it currently exists, originated from environmental communication." Flor delineates six essential elements of environmental communication: comprehension of ecological principles; recognition of cultural influences; adeptness in networking; efficacy in employing media for social agenda development; dedication to environmental ethics; and competencies in conflict resolution, mediation, and arbitration [2]. Cox categorises the domain of environmental communication into seven key areas: environmental rhetoric and discourse, media and environmental journalism, public engagement in environmental decision-making, social marketing and advocacy initiatives, environmental collaboration and conflict resolution, risk communication, and portrayals of nature in popular culture and green marketing [2].

Researchers see environmental communication as both symbolic and concrete [3]. Humans react to their surroundings and develop beliefs based on their perceptions. The phrase "call to action" is significant in environmental communication, as it pertains to matters like climate change, endangered species, and pollution [4]. Thus, environmental communication is crucial to sustainability research by converting knowledge into action. Environmental communication, focused on everyday speech and cooperation, has a deep understanding of public discussion over environmental policy [5]. Sustainability science requires cooperation among stakeholders, necessitating efficient communication to enable sustainable change. Cox is a notable authority in environmental communication and its importance in the public sphere [2]. Cox examines the importance of environmental communication and its impact on government, advocacy initiatives, media, and environmental movements [6]. Cox overlooked the importance of environmental communication in the public sphere, especially the roles of visual and aural communication, electronic and digital media, and, specifically, popular culture mostly influenced by New Media.

1.2 History of Environmental Communication

Humans have used a variety of communication techniques to interact with one another and their surroundings throughout history [7]. Some of these techniques include simple gestures and verbal cues to more advanced forms of communication such as writing, printing, and digital technologies. As society modernised, the industrial complexes and factories on which modernity was founded posed a significant threat to the environment through air, water, and soil pollution, chemical discharge, resource depletion, and other forms of environmental degradation. In response to growing concerns about the negative impact of industrialisation, natural disasters, the effects of globalisation on the environment, and the need to conserve and preserve the environment, the field of environmental communication emerged in the United States in the 1970s deviating from standard rhetorical theory [8]. As a consequence, organisations were established with the goal of educating the public about environmental problems and encouraging them to take action to save the environment.

The publication of Rachel Carson's *Silent Spring* in 1962 marked a significant milestone in the evolution of environmental communication [9]. This article highlighted the detrimental effects of pesticides on the ecosystem and aided in fostering awareness about the need of environmental protection. Carson's compelling writing and convincing arguments encouraged many people to take action and laid the groundwork for future environmental communicators to advocate for change [10].

The inception of Earth Day in 1970 was another pivotal factor in the development of environmental communication [11]. This annual event, now internationally recognised, heightened awareness of environmental issues and encouraged everyone to reflect on the impact of their actions on the planet. Earth Day functioned as a platform for environmental communicators to share information, mobilise communities, and promote legislative changes to improve environmental well-being [12].

With the rise of environmental journalism, documentaries, and advocacy campaigns, establishing an environmental communication magazine in 2007, environmental communication expanded and became more varied [13]. As awareness of environmental issues increased and the need for communication about these issues became evident, this led to the term "environmental communication" being coined in the 1970s [14]. With the founding of international organisations such as Greenpeace and the United Nations Environment Programme, the 1980s and 1990s saw a growing recognition of the global nature of environmental issues [15].

Environmental communication has undergone a revolution in recent years due to the emergence of digital media, which has supplemented traditional media and made it easier for people to communicate and exchange information about environmental issues. Promoting environmental causes and reaching a larger audience have been made possible in large part by the media. In order to promote action and increase public awareness on problems such as climate change, deforestation, and biodiversity loss, environmental organisations and activists have also learned how to use the media [16].

The need for efficient environmental communication will only increase as society progresses. The health of our planet and the welfare of future generations are seriously threatened by climate change, pollution, and habitat destruction. Communication organisations must never stop informing, educating, and provoking dialogue about environmental issues and the significance of taking action to safeguard the planet we all share. The transmission of knowledge, concepts, and messages about environmental issues is the fundamental component of environmental communication. There are numerous ways to do this, including through digital and traditional media [17]. Raising awareness, igniting action, and fostering a deeper comprehension of the intricate relationships between society and the environment are the objectives of environmental communication.

1.2.1 Factors That Led to the Emergency of Environmental Communication

The emergence of environmental communication can be attributed to a combination of factors, including the increasing severity of environmental problems, advances in technology, and changing societal values. As the planet faces unprecedented environmental challenges, it became more important than ever to communicate the importance of protecting the environment and promoting sustainable practices. By raising awareness, engaging with the public, and advocating for policy change, environmental communication can help create a more sustainable future for our planet.

1.2.1.1 Industrialisation

As industrialisation grew in the middle of the twentieth century, environmental issues became more urgent [14]. Natural resource depletion and soil, water, and air pollution are direct effects of industrialisation. It therefore became crucial to convey the detrimental effects of these activities on the environment as industries grew and used resources at an alarming rate. As a result, environmental communication emerged to draw attention to the negative effects of industrial practices and to promote the use of environmentally friendly substitutes [18].

Natural disasters like climate change and biodiversity loss are a result of industrialisation. Effective communication strategies were necessary to raise public awareness of the pressing need for action in response to these global challenges. Hence, environmental communication became crucial for galvanising support for environmental conservation initiatives and influencing policy choices that can lessen the environmental damage caused by industrial operations [18] (Fig. 1.1).

As a result of industrialisation, industries attempted to minimise their negative environmental effects, resulting in a proliferation of misinformation and greenwashing. This also had an impact on the development of environmental communication, which is now a vital tool for refuting false allegations and promoting accountability and transparency in business practices. Environmental communication facilitated the empowerment of individuals and communities to make sustainable decisions by disseminating precise and trustworthy information [5].

Furthermore, as a result of industrialisation, there was a growing market for sustainable goods and services, which led companies to use green marketing techniques to draw in eco-aware customers [2]. As a result, environmental communication emerged to establish credibility and trust with customers by outlining the advantages that products and services have for the environment.

Fig. 1.1 Industrialisation. Courtesy of Wikimedia Commons (2025)

The transition to a more sustainable economy was aided by environmental communication, which encouraged openness and consumer discussion.

1.2.1.2 Globalisation

The environment was impacted by globalisation due to an increase in pollution [7]. Globalisation refers to the interconnectedness of countries and civilisations via the exchange of goods, services, information, ideas, technology, and culture. The interconnectedness is facilitated by developments in technology, communication, transportation, and commerce, enabling corporations, governments, and individuals to interact on a worldwide scale [9]. The swift growth of global trade and manufacturing led to increased emissions of greenhouse gases, contamination of air and water, and generation of waste. This was a major factor in ecosystem destruction, biodiversity decline, and global warming [2]. As a result, environmental communication became increasingly important in informing the public about the negative effects of pollution and motivating people and companies to switch to more environmentally friendly practices.

The depletion of natural resources is another way that globalisation affected the environment. Overuse of natural resources, deforestation, and habitat destruction resulted from the increased demand for energy and raw materials to support economic growth [8]. This endangered not just wildlife and ecosystems but also future generations' access to vital resources. As a result, environmental communication was created to support sustainable development, resource conservation, and responsible consumption in order to help address this problem (Fig. 1.2).

In addition, globalisation made it easier for dangerous substances and chemicals to cross international borders, thus endangering both the environment and human health. Pollutants like pesticides, heavy metals, and industrial

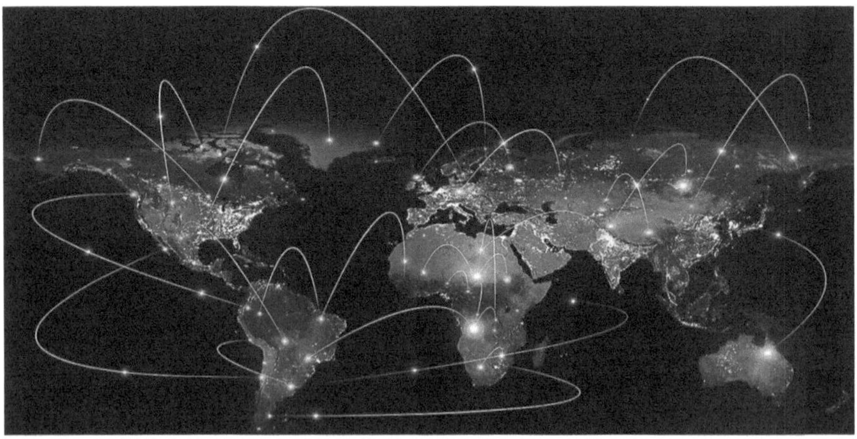

Fig. 1.2 Globalisation. Artificial Intelligence Generated (2025)

chemicals widely contaminated air, water, and soil as a result of the global movement of goods and materials [8]. This had detrimental effects on ecosystems, wildlife, and human populations in addition to increasing the prevalence of diseases linked to pollution and having a negative impact on health. Therefore, environmental communication was developed to educate the public about the dangers of pollution exposure and to promote laws that give public health and environmental protection top priority.

Research indicates that urbanisation accelerated due to globalisation, with cities and infrastructure expanding quickly at the expense of biodiversity and natural habitats [9]. The conversion of wetlands, forests, and other natural areas for industrial, residential, and agricultural uses caused ecosystem disruption, habitat fragmentation, and biodiversity loss. The environment suffered greatly as a result, losing important ecosystem services, becoming more vulnerable to natural disasters, and becoming less resilient to climate change. In order to lessen these adverse effects and protect biodiversity for future generations, environmental communication was required to support conservation initiatives, habitat restoration, and sustainable urban planning techniques.

1.2.1.3 Natural Disasters

Natural disasters brought to light how closely human activity and the environment were related. People were seeing first-hand the catastrophic effects of climate change and environmental degradation on the planet through natural disasters like hurricanes and wildfires. As a result, public education about the value of sustainable practices and the part that each person can play in reducing these risks received more attention. The goal of environmental communication was to present these concerns in a way that people from all walks of life could understand and find interesting [2].

Environmental issues had a sense of urgency as a result of natural disasters. Communities were more cognisant of the necessity of efficient communication and teamwork in order to respond to and recover from disasters when they were immediately threatened by one [4]. The purpose of environmental communication was to spread knowledge about crisis management, evacuation protocols, and emergency preparedness [6]. Environmental communication was therefore developed to help make sure that communities were better prepared to deal with the effects of natural disasters by encouraging timely and clear messaging.

Research indicates that natural disasters spurred a global dialogue about the importance of environmental stewardship and sustainable development [6]. Following natural catastrophes such as hurricanes, floods, and earthquakes, there was frequently a renewed emphasis on rebuilding in a way that is more resistant to recurrence. Thus, environmental communication was developed to promote laws and procedures that give the environment's health and the welfare of future generations top priority. Environmental communication was devised to promote responsible decision-making and sustainable solutions by

Fig. 1.3 Natural disaster. Courtesy of Wikimedia Commons (2025)

increasing public awareness of the long-term effects of environmental degradation (Fig. 1.3).

Furthermore, the significance of cooperation and community involvement in tackling environmental issues was highlighted by natural calamities [9]. Governments, groups, and individuals were expected to collaborate in the event of a disaster in order to support vulnerable populations and coordinate response efforts. In order to foster discussion, information exchange, and group action among stakeholders, environmental communication was established [10]. Environmental communication was developed to increase community resilience to environmental threats by promoting participation from all societal segments and a sense of shared responsibility.

1.3 Principles of Effective Environmental Communication

In order to promote sustainability, increase awareness, and encourage positive change in the direction of a healthier planet, it is imperative that environmental communication adheres to certain principles. It is possible to motivate people to take action, instil a sense of responsibility, and improve the future for future generations by adhering to environmental communication principles. Unquestionably, communication has great power, and by using it wisely, everyone can contribute to environmental preservation and the development of a more sustainable global community. The principles of effective communication include understanding the audience, using clear and simple language, using credible sources, making use of multiple channels, promoting bidirectional dialogue, tailoring messaging, using visual aids, using culturally sensitive communication, empowering messages, repetitive and consistent messaging, evaluating and adapting, interactive communication, sustainable communication, entertainment education, and social change.

1.3.1 Understanding the Audience

To communicate effectively, communicators should first understand their audience's values, beliefs, and attitudes to craft messages that resonate with them. When information is pertinent to people directly, they are more likely to pay attention to it and act upon it [4]. Environmental communicators should determine the goals, concerns, and motivations of their target audience and adjust their messaging by performing audience research and analysis. For example, parents who are worried about their children's health can be more receptive to a message about the health benefits of clean air, whereas business owners can find greater interest in a message about the financial advantages of green energy.

Moreover, knowing their audience makes it easier for communicators to select the best formats and channels for information distribution. Information can be consumed in a variety of ways by different audiences, whether it be through direct mail, social media, traditional media, or local events. Environmental communicators can enhance the effectiveness and reach of their messages by being aware of the preferences of their target audience. For instance, whereas older adults might prefer printed materials or in-person presentations, young adults might be more inclined to interacting with environmental messaging on social media sites like Instagram and TikTok.

Communicators can effectively address scepticism, misinformation, and misconceptions when they have a thorough understanding of their audience. They can anticipate potential obstacles and objections and proactively address them in their messaging by understanding the attitudes and beliefs of their audience [3]. Besides, communicators can counter misinformation and establish trust, for instance, by presenting convincing evidence and real-world examples to an audience that is sceptical of the science behind climate change.

In addition, a sense of empathy and a connection between communicators and their audience are fostered by understanding the audience [11]. Environmental communicators can create rapport and trust with their audience by taking the time to learn about their viewpoints, worries, and experiences. These qualities are crucial for inspiring behaviour change. Audiences are more likely to be open to messages and take action to protect the environment when they feel heard and respected.

1.3.2 *Use Clear and Simple Language*

Using plain language facilitates communication between specialists and the general public [9]. Understanding many environmental issues fully requires specialised knowledge due to their high level of technicality. Nonetheless, communicators can make sure that everyone, regardless of background or education level, can comprehend the significance of environmental conservation by dissecting technical jargon and using simple language. In order to promote a sense of solidarity and cooperation in the battle against climate change and other environmental threats, this inclusivity is essential. The use of plain language makes it easier to fight confusion and false information [5]. In the current digital era, false information spreads quickly, making it challenging for the general public to distinguish between fact and fiction. Communicators can assist in dispelling myths and misconceptions by providing information in an understandable and direct manner, enabling people to make decisions about how their actions will affect the environment. This clarity fosters a sense of accountability and responsibility for our actions in addition to stimulating critical thinking.

Additionally, action is encouraged, and engagement is increased by using plain language [2]. When environmental communication strikes a personal chord with its audience, it works best. Through the use of relatable and easily understood language, environmental communication can arouse feelings and motivate people to take constructive action in the direction of environmental conservation. When it comes to cutting down on plastic waste, saving water, or promoting sustainable practices, effective communication can inspire people to make significant life changes. Trust and transparency are promoted by plain and straightforward language [5]. In a time of increased mistrust and cynicism, environmental communicators must be forthright and honest about the difficulties people face. Environmental communicators can gain the public's trust by speaking clearly and transparently, which can increase the credibility of environmental communication. The long-term connections and collaborations required to make significant advancements in environmental sustainability depend on this trust.

1.3.3 *Credible Sources*

It is easier to demonstrate the authority and legitimacy of the information being shared when reliable sources are used [5]. People are more likely to believe and trust information when they can verify that it is coming from reliable sources, such as government agencies, reputable academic institutions, or well-known authorities in the field. Since trust is the basis for swaying attitudes and actions regarding environmental issues, it is essential for effective communication. Reputable sources give the arguments and assertions made in environmental communication a strong basis. Communication initiatives can provide well-researched and verified data and evidence to support their claims by citing reliable sources. This helps to persuade sceptics and doubters of the veracity of arguments while also reinforcing message.

In addition, spreading false information and misinformation can be prevented by using reliable sources. In the current digital era, social media and other online channels make it far too simple for inaccurate or misleading information to proliferate rapidly. Environmental communication stops the spread of false information and gives the audience trustworthy, accurate information by relying on reputable sources. Research indicates that establishing trustworthiness is crucial for creating enduring bonds with viewers [7]. People are more likely to think highly of the organisation as a trustworthy and dependable information source if they observe that the organisation regularly cites credible sources and delivers accurate information. By doing this, the organisation can increase audience loyalty and credibility, which will boost participation in and support for environmental initiatives.

1.3.4 Making Use of Multiple Channels

Reaching a larger and more varied audience is one of the main advantages of using multiple communication channels. Environmental communicators can reach people of all ages, backgrounds, and locations by combining traditional media, social media, websites, email newsletters, and live events. This extensive outreach is crucial for bringing environmental issues to people's attention and motivating them to take action to safeguard the environment.

Since environmental issues are frequently intricate and multidimensional, communication strategies must be nuanced [7]. Environmental communicators can convey information in a variety of formats, including written articles, infographics, videos, and interactive websites, by utilising a variety of communication channels. This variety of content facilitates improved message comprehension and increases the audience's ability to understand complex ideas. Additionally, by repeating and reinforcing important messages across multiple channels, comprehension and retention are further strengthened.

In addition to providing information, effective environmental communication also aims to engage the audience and promote interaction [7]. Environmental communicators can foster opportunities for reciprocal dialogue, feedback, and cooperation by utilising a variety of communication channels. In particular, social media platforms provide a forum for people to connect with other environmentalists who share their interests and ask questions. This kind of involvement fosters a feeling of community and gives people the confidence to take significant action in response to environmental issues.

Since people are inundated with information from a multitude of sources in the modern digital age, environmental communicators must establish credibility and cultivate trust with their audience. Environmental communication organisations can exhibit transparency, consistency, and dependability in their messaging by utilising a variety of communication channels. People tend to view a message as more reliable and trustworthy when they see the same information presented in a variety of ways. Credibility is essential for motivating people to support the organisation's environmental protection efforts and for fostering trust in the organisation's mission.

1.3.5 Promote Bidirectional Dialogue

Information can be exchanged between parties more meaningfully when two-way communication is implemented. By promoting a conversation instead of a monologue, all parties are able to express their viewpoints, thoughts, and worries. This promotes communication and cooperation, which improves decision-making and the results of environmental projects. The advancement of environmental goals can be hampered by miscommunication, misunderstandings, and conflicts that can arise from a lack of two-way communication [8].

Among stakeholders, encouraging two-way communication contributes to credibility and trust-building. People are more likely to be open to messages when they feel heard and appreciated. This fosters a climate that is favourable to collaboration and partnership, both of which are necessary to handle challenging environmental issues. When it comes to inspiring people and institutions to take up sustainability and conservation-related activities, trust is a crucial component (Fig. 1.4).

A two-way communication makes it possible to receive input and feedback from a variety of angles. Given the multitude of stakeholders involved in environmental communication, each with varying interests and priorities, this is particularly crucial. Environmental initiatives can become more inclusive, equitable, and successful by hearing different points of view and incorporating feedback into the decision-making process. By doing this, the case for sustainable behaviours that benefit both people and the environment is strengthened. Besides, encouraging two-way communication facilitates continued interaction and discussion with stakeholders. Since environmental issues are always changing, it is critical to keep lines of communication open in order to adjust to new situations and challenges. Establishing a discourse and cooperation culture can help organisations become more resilient and adaptable when dealing with environmental problems. This fosters a sense of shared accountability and ownership among participants, increasing dedication and

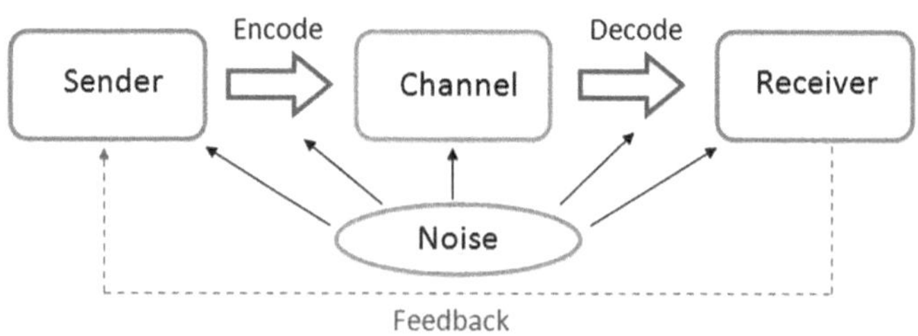

Fig. 1.4 Bidirectional dialogue. Author (2025)

increasing the likelihood that environmental conservation initiatives will be successful in the long run.

1.3.6 Tailored Messaging

Research shows that personalised communication enables more effective interaction with the intended audience [9]. The audience is more likely to pay attention and take in the information when content is tailored to suit their interests, values, and worldview. To effectively convey the message, it might be more effective to use graphics and images if the audience is more visually oriented. Environmental communicators can improve the chances that their message will be heard and comprehended by addressing the wants and needs of their audience.

Information can be made more relatable and relevant to the audience by using tailored messaging [9]. If someone can clearly see how a problem affects them or their community, they are more likely to take action on it. Environmental communicators can make information more compelling and personal by developing messages that specifically address the interests and concerns of their target audience. For instance, if the audience's main concern is air quality, environmental communication initiative can encourage them to take action to improve the quality of the air in their community by customising messages to emphasise the negative effects of pollution on respiratory health. A messaging that is specifically tailored to the audience can increase credibility and trust. The audience is more likely to regard a communicator as knowledgeable and reliable when the messages are pertinent and resonate with them. Environmental communicators can build trust and establish a stronger connection by exhibiting a comprehension of the needs and values of their audience.

In order to inspire the audience to act and take part in environmental initiatives, this trust is crucial. Additionally, more precise targeting of particular groups or demographics is made possible by tailored messaging. Regarding environmental issues, different audiences might have differing levels of knowledge, attitudes, and behaviours. Environmental communicators can more effectively adapt their communication strategies to each group's needs by crafting messages that specifically address the characteristics of each audience segment. For instance, communications aimed at policymakers might demonstrate the financial advantages of sustainable practices, while communications aimed at young adults might highlight the effects of climate change on future generations.

1.3.7 Using Visual Aids

Scholars explain that visual aids facilitate the more efficient and clear communication of information. Environmental communicators can simplify complex ideas for their audience by using pictures, graphs, charts, and videos. This makes it possible for people to comprehend the current environmental problems and suggested solutions better. For example, a graph that illustrates the rise in carbon emissions over time is far easier to comprehend and has far more impact than simply stating the numbers.

Besides, visual aids have the power to arouse feelings in viewers and establish a rapport [11]. Environmental problems are frequently impersonal and unrelated to people's everyday experiences. Communicators can personalise and relatable these issues by using poignant images or videos. People can be inspired to take action against climate change, for example, by seeing an image of a polar bear that is starving on a melting ice cap.

Studies also indicate that using visual aids improves memory recall [5]. When information is presented visually as opposed to presenting it textually, people are more likely to remember it. Environmental advocates can improve the chance that their message will be remembered by their audience by using visual elements in their communication. This is especially crucial when attempting to influence behaviour or motivate environmental action.

Visual aids can also assist in reaching a larger audience. Not all people learn best when they read or listen to material [3]. Communicators can engage a wider audience and accommodate various learning styles by utilising visual elements. This is essential for raising public awareness of environmental problems and enlisting support for remedies.

1.3.8 Culturally Sensitivity Communication

Diverse cultural backgrounds give rise to distinct viewpoints regarding environmental matters [5]. Environmental communicators can make their messages more relatable to their audience by taking the time to learn about the cultural values and beliefs of various communities. This promotes inclusivity and demonstrates respect for different points of view, which increases acceptance of and participation in environmental initiatives.

Communication that is sensitive to cultural differences shows a dedication to comprehending and honouring the needs and worries of various communities. This increases audiences' sense of trust and credibility, which raises the possibility that they will be open to environmental messages and take action to safeguard the environment. Culturally sensitive communication can help close the gap between environmental organisations and the communities they serve in a world where trust in institutions is eroding.

Besides, the capacity to captivate and inspire audiences to act is essential for effective environmental communication. Environmental communicators can adapt their messages to resonate with the values, beliefs, and experiences of diverse communities by using culturally sensitive communication. This increases the communication's impact and relatability, raising the possibility that listeners will internalise the message and take significant action to safeguard the environment.

In addition, all people are impacted by environmental issues, irrespective of their cultural background. Environmental communicators can make sure that a wide range of audiences hear and are engaged by their messages by employing culturally sensitive communication. This encourages inclusivity and diversity within the environmental movement, resulting in a more equitable and inclusive strategy for dealing with environmental issues. Environmental communicators can encourage a sense of shared responsibility for safeguarding the environment and creating a sustainable future for all by embracing cultural diversity and promoting inclusivity.

1.3.9 Empowering Messages

Messages that are inspiring and empowering have potential to give people a sense of agency and empowerment. These messages inspire people to take responsibility for their actions and make a difference by emphasising the positive effects of individual actions on the environment. Individuals are more likely to act and adopt sustainable practices when they believe they have the power to change things. In fact, positive and inspiring messages help people feel hopeful and optimistic. These messages serve as a reminder that constructive change is achievable in a world where environmental issues can frequently seem overwhelming and unsolvable. Through the presentation of accomplishments and advancements, these messages encourage people to have faith in the potential for a sustainable future and the strength of group efforts [9].

Besides, messages that are inspiring and uplifting contribute to the development of a sense of unity and community. These messages inspire people to band together and strive towards a common objective by highlighting the shared responsibility that each of us has towards the environment. People are more likely to remain inspired and dedicated to changing the world when they perceive themselves as part of a greater cause and receiving support from a group of like-minded people.

In addition, empowering and motivating messages can help people overcome feelings of helplessness and apathy, according to studies [5]. These messages serve to remind people that their actions matter and that they have the power to bring about positive change in a world where environmental issues are frequently minimised or disregarded. People are encouraged to take proactive steps towards a more sustainable future by these messages, which emphasise the significance of individual actions and the cascading effects that these actions can have.

1.3.10 Repetitive and Consistency Messaging

Scholars explain that public perception of the significance of environmental issues is bolstered by repeated and consistent messaging [13]. When a message is repeated through different channels, people are more likely to absorb its meaning and pay attention to it. People are more likely to take action to address environmental issues in their daily lives when they hear these repeated messages because they help keep these issues at the forefront of their minds.

According to research, messaging that is repeated and consistent with the audience can help establish credibility and trust [3]. When individuals witness a consistent message being disseminated across various platforms, they are inclined to accept its veracity and have faith in the information's origin. This trust is essential when it comes to starting discussions about environmental issues and motivating people to change their behaviour for the better. To cultivate a sense of urgency about environmental issues, the use of persistent and consistent messaging can be deployed. Organisations can inspire people to take action and change the world by consistently educating people about the value of environmental protection and the negative effects of inaction. People can be more supportive of environmental programmes and laws as a result of this sense of urgency, and they can also be motivated to make personal adjustments in order to lessen their own environmental impact. Thus spreading recurring and consistent messaging can foster a feeling of community among persons who have similar environmental concerns. People who are passionate about environmental protection can feel more connected to one another if they hear the same message over and over. When individuals and organisations are working towards the same objective of protecting the environment for future generations, a stronger sense of community can foster greater engagement and collaboration.

1.3.11 Evaluating and Adapting

Research shows that the process of monitoring and adapting environmental communication can help messages reach and engage with the intended audience [5]. Because not everyone processes and absorbs information in the same manner, it is vital to regularly assess the effectiveness of communication efforts and make necessary changes to establish ties with the audience. Evaluating and adapting environmental communication can improve the effect and engagement of environmental communication messages by changing and reacting to feedback.

In a rapidly changing environment, individuals may stay informed by evaluating and updating communication strategies. The fact that environmental concerns are dynamic and always evolving, communication methods must evolve to capitalise on these new opportunities and obstacles. Environmental communication campaigns can ensure that they are addressing the most pressing environmental issues while also providing their audience with accurate and up-to-date information by consistently assessing the success of communication strategies.

It is possible to increase the validity and reliability of environmental communication messages by evaluating and adjusting communication tactics. Establishing trust with the audience is critical in environmental communication because it encourages action and influences positive behaviour. By demonstrating commitment to continuous improvement and transparency in their communication efforts, environmental communication initiatives can strengthen ties with their audiences.

The process of monitoring and adapting environmental communication can help identify what works and what does not, which can help maximise the effect of their efforts. By carefully analysing and learning from previous experiences, environmental communication projects can better concentrate their communication efforts and allocate resources. Using this iterative review and adaption process, environmental communication campaigns can enhance their messages and, as a result, increase greater awareness and action on environmental challenges.

1.3.12 *Interactive Communication*

Studies indicate that people who engage in interactive communication feel more empowered and in control of their lives [4]. Engaging stakeholders in the communication process can increase their likelihood of feeling committed to the environmental cause and inspiring them to take action. Individuals are more likely to contribute to sustainable solutions and alter their behaviour for the better when they feel heard and respected in the conversation. Interactive communication makes it possible for different viewpoints and ideas to be shared. Different stakeholders can share their knowledge, experiences, and concerns through open dialogue and discussions, which will result in a more thorough understanding of environmental issues. By addressing potential conflicts, finding creative solutions, and fostering consensus among diverse groups, this collaborative approach eventually results in more successful environmental initiatives (Fig. 1.5).

Besides, interactive communication also encourages accountability and transparency. Organisations and decision-makers can show their commitment to environmental responsibility and cultivate trust with stakeholders by promoting open communication. The organisation's reputation is improved, credibility is established, and goodwill is fostered—all of which contribute to increased support for the organisation's environmental initiatives. By facilitating evaluation and feedback, interactive communication enables organisations to continuously improve their environmental communication strategies. Organisations can evaluate the success of their messaging, spot areas for development, and make necessary adjustments by asking stakeholders for feedback and tracking their answers. Organisations can adjust to shifting conditions, take proactive measures to address issues, and make sure that their environmental initiatives are meaningful and resonate with their target audience thanks to this feedback loop [9].

1.3.13 *Sustainable Communication*

Research shows that delivering environmental communication messages in a way that respects the environment and does not cause more harm is guaranteed

Fig. 1.5 Interactive communication. Courtesy of Freepik (2025)

by sustainable communication [11]. It is important for environmental communication initiatives to have an understanding of the resources and energy used in creating and disseminating messages in a world where information is continuously thrown at people through a variety of channels. Utilising recycled materials, cutting back on waste, and reducing carbon footprint are just a few examples of sustainable practices which environmental communication initiatives could incorporate into their communication campaigns to demonstrate that these initiatives are doing their part to preserve the environment.

Addition, given an audience that is eager to listen, sustainable communication can aid in establishing credibility and trust. The public is growing more sceptical about the messages they hear in an age of greenwashing and exaggerated claims of environmental stewardship. Environmental communication initiatives can demonstrate that communicators of these messages are dedicated to not just talking the talk but also walking the talk by making sure that their communication efforts are in line with sustainable principles. By doing so, environmental communication programmes can increase audience trust and make sure that their messages are understood and taken to heart. Environmental communication initiatives can reach a larger audience and magnify their message through sustainable communication. Such initiatives can show prospective partners and collaborators that the stakeholders are dedicated to having a positive environmental impact by implementing sustainable practices in their communication efforts. This can draw groups and individuals who share environmental communication initiatives' beliefs and are prepared to collaborate in order to spread their message and reach a larger audience. Establishing a network of partners who share the commitment to sustainability can help environmental communication programmes to accomplish more and make a bigger impact with their communication efforts.

Lastly, inspiring action and fostering a culture of environmental stewardship are two benefits of sustainable communication. Environmental communication initiatives can show their audience that living and working in an environmentally responsible and respectful manner is feasible by incorporating sustainable practices into communication campaigns. This has the potential to motivate people to take personal responsibility and enable them to have a beneficial influence on the environment. Environmental communication initiatives can initiate significant change by setting an example and demonstrating that sustainable communication is not only feasible but also successful.

1.3.14 Entertainment Education

Compared to traditional educational methods, entertainment education can reach a larger audience [4]. When information is presented in an interesting and captivating way, people are more likely to interact with it and remember it. Environmental messages can be conveyed in an engaging and approachable way by fusing entertainment components like humour, storytelling, and images with instructional content. This can draw in a wider audience, including people who might not normally look for environmental information. In essence, entertainment education has the power to arouse feelings in viewers and establish a bond with them. Apathy or disinterest can sometimes result from environmental issues feeling overwhelming and abstract. Entertainment education, however, has the power to inspire empathy, concern, and action by delivering information in a way that appeals to people's emotions. For instance, a compelling documentary or movie that shares the experiences of people impacted by climate change can encourage viewers to reconsider their own actions and lend support to environmental causes (Fig. 1.6).

A platform for encouraging constructive behavioural change can be found in entertainment education. In addition to increasing awareness, environmental communication aims to persuade individuals to incorporate sustainable practices into their daily lives. Audiences can be encouraged to make environmentally conscious decisions without feeling lectured or coerced by incorporating environmental messages into entertaining media like TV shows, movies, or video games. A plan that emphasises the advantages of carpooling or recycling, for example, can gently persuade viewers to take up these environmentally beneficial practices.

Conversations about environmental issues can be facilitated and barriers can be broken down with the aid of entertainment education. A common ground for conversation and debate can be established through entertainment, which presents information in a non-judgemental and nonthreatening manner. This can be especially useful when trying to reach audiences with varying backgrounds or points of view on environmental issues. Through the promotion of critical thinking and open communication, entertainment education can enable

Fig. 1.6 Entertainment education. Courtesy of Wikimedia Commons (2025)

people to take responsibility for environmental problems and actively participate in finding solutions.

1.3.15 Social Change

Evidence shows that social change facilitates the transition of cultural norms and values towards a future that is more sustainable [11]. By interacting with communities and emphasising the value of environmental conservation, environmental communication can start a chain reaction that encourages more people and organisations to adopt environmentally friendly practices. This can involve anything from promoting renewable energy sources to lowering the use of single-use plastics. These adjustments can become deeply embedded in people's culture as more people adopt them, which can have lasting effect on the environment.

Local, national, and international policy can also be influenced by social change. Communities can be organised and mobilised around environmental issues to put pressure on legislators to enact laws that safeguard the environment. This can involve taking steps to promote sustainable agriculture, save natural habitats, and lower carbon emissions. These policies can make a big difference for the future generations and the planet's health if they received enough support from the general public.

Besides, addressing environmental injustices and inequalities can be aided by social change. Environmental communication can strive towards a more just and equitable world by elevating the voices of marginalised communities that suffer disproportionately from environmental degradation. These communities can demand equitable access to clean water and air, as well as defence against pollution and environmental dangers, if social change takes place. Environmental communication can create a society that is more inclusive and sustainable by fighting for environmental justice.

In addition, social change also fosters a sense of empowerment and group responsibility. When people work together to solve environmental issues, they can accomplish far more than any one person could on their own. Social change has the power to motivate individuals to take action and improve the world by promoting a sense of belonging and solidarity. Environmental communication can help people to build a more resilient and sustainable future for themselves and future generations by working together.

1.4 The Importance of Environmental Communication

Environmental communication is crucial to educating and enlightening the public about the state of the planet and the actions that need to be taken to preserve it for future generations. It is becoming more and more crucial to achieve effective communication about global issues such as pollution, deforestation, climate change, and loss of biodiversity.

1.4.1 Empowering Marginalised Communities

Drawing attention to the particular environmental issues that marginalised communities face is one way that environmental communication empowers these marginalised communities. Because of systemic injustices and discrimination, these communities are frequently disproportionately impacted by environmental problems such as pollution, climate change, and other environmental problems. Marginalised groups can guarantee that their voices are heard and that awareness of these issues is increased by using powerful communication techniques like online platforms, community meetings, and media campaigns. By drawing attention to the environmental injustices that these communities experience, we can mobilise allies and put pressure on policymakers to act [11].

Environmental communication not only increases awareness but gives marginalised communities a forum for group action. Through facilitating connections between individuals confronting comparable environmental issues, these channels of communication can enable communities to unite, organise, and champion their rights. Through lobbying, petitions, or grassroots organisation, environmental communication can support marginalised groups in raising their voices and demanding answers to the problems they face. Acting in concert, these communities can bring about positive change for themselves and the next generation while holding those in positions of authority accountable. By encouraging marginalised communities to participate in decisions that impact their lives, environmental communication can strengthen these communities. Communications channels can guarantee that these communities are included in discussions about policies and initiatives that affect their environments by offering easily accessible information, resources, and opportunities for engagement. Through online surveys, public hearings, or town hall meetings, environmental communication can give marginalised people a voice and allow them to affect the decisions that are made about the environment. These communities can make sure that their needs are met, their rights are upheld, and their voices are heard by taking part in these processes.

1.4.2 *Enhance Sustainability Behaviour*

Environmental communication plays a vital role in educating the public about the value of sustainability and environmental conservation. A large audience is reached by messages about resource depletion, biodiversity loss, and climate change through a variety of platforms, including social media, educational initiatives, and public campaigns. People learn more about their role in promoting sustainability when they are informed about the effects of human activity on the environment. People who possess this knowledge are better equipped to make responsible decisions and protect the environment [13]. Environmental communication encourages discussion and participation from a range of stakeholders, such as the public, businesses, non-governmental organisations, and policymakers. Communication can result in the development of common objectives and solutions for environmental problems by establishing forums for debate and cooperation. Public consultations on environmental policies, for example, give people a chance to express their ideas and participate in the decision-making process.

Partnerships between companies and environmental groups can also result in creative projects that advance sustainable practices. Effective communication allows stakeholders to collaborate on shared goals and bring about long-lasting change.

People can be inspired to take action and be motivated to adopt sustainable behaviours in their daily lives through environmental communication. Communication campaigns have the power to emotionally connect people and motivate them to take action by utilising gripping narratives, eye-catching imagery, and calls to action. Campaigns that support energy efficiency, recycling, and sustainable transportation, for instance, can persuade people to adopt eco-friendly habits. Role models' testimonies and personal accounts can also demonstrate the impact that one person's actions can have and encourage others to take similar action.

1.4.3 *Promoting Partnerships*

Public education and awareness of the value of environmental conservation are fostered by environmental communication [4]. Through a variety of platforms, including websites, social media, and public gatherings, environmental communication can spread knowledge and motivate people to take action. People are more likely to support initiatives that promote sustainability and alter their behaviour when they are aware of how their actions affect the environment [18]. By fostering a shared understanding and sense of accountability, this shared knowledge strengthens the bonds between various stakeholders.

Intergroup communication about the environment fosters discussion and cooperation despite differences in goals and viewpoints. Environmental communication can bring people, organisations, and governments together to share best practices, find common ground, and exchange ideas by establishing a forum for discussion and debate. Stakeholders can pool their resources and expertise to achieve greater impact when they can effectively communicate and work towards a common goal. Environmental communication can create trust among collaborators and bridge the gap between divergent interests by promoting partnerships through honest and open communication.

Furthermore, through involving stakeholders and creating an alliance network, environmental communication can mobilise support and resources for environmental initiatives. Environmental communication can motivate action and stimulate funding for environmentally friendly technology and conservation initiatives by highlighting the advantages and possibilities of sustainable practices. People are more likely to support and promote change when they identify with a cause and recognise the importance of their contributions [7]. Environmental communication can multiply people's effect and produce a far-reaching ripple effect by building a community of environmental champions through efficient communication.

1.4.4 Bridging the Gap

Policymakers and the general public can benefit greatly from scientific research and findings being shared through environmental communication. Researchers, data gatherers, policymakers, and scientists can have a significant impact on public awareness campaigns and environmental policy. But in the absence of clear communication, decision-makers and the general public might not be able to access or understand this important information. Environmental communicators can effectively convey scientific research through reports, articles, and presentations, which can help translate complex scientific findings into language that the public can easily understand and use to inform policy decisions and raise awareness of urgent environmental issues [19].

Besides, collaboration and discussion between scientists, policymakers, and the general public are made easier by environmental communication. Environmental communication facilitates the sharing of thoughts, viewpoints, and solutions to environmental problems by giving stakeholders a forum for candid and transparent dialogue. Environmental communicators can facilitate meaningful conversations that foster mutual understanding and cooperation between policymakers, scientists, and the public through various forums, including public hearings, stakeholder meetings, and social media platforms.

By developing relationships, establishing trust, and identifying points of agreement on environmental issues, this conversation can aid in bridging the divide between these important stakeholders [7].

In addition, environmental communication acts as a spur to action and transformation. Environmental communication has the power to inspire legislators, scientists, and the general public to take decisive action by increasing awareness, gaining support, and igniting action in response to environmental concerns. Environmental communicators can instil a sense of urgency and give people the power to make decisions that will lead to a more sustainable future through campaigns, advocacy work, and storytelling. Environmental communication can spur constructive action that improves the environment and society at large, whether it be through supporting conservation efforts, arguing for legislative changes, or encouraging sustainable practices [20].

1.4.5 Foster Public Participation

Research shows that environmental issues and their effects on the environment are brought to light through environmental communication [9]. Environmental communicators can inform the public about the seriousness of environmental issues such as pollution, deforestation, climate change, and habitat destruction by skilfully presenting scientific research, data, and real-world examples. People are more likely to feel inspired to take action and make changes in their own lives to help protect the environment when they are aware of the threats facing our planet. Environmental communication offers forums for public discussion and involvement. Environmental communicators can enable people to express their thoughts, worries, and solutions for resolving environmental issues by establishing forums for conversations, debates, and community involvement. Environmental communication can improve the efficacy of environmental policies and initiatives by taking into account public feedback and listening to a variety of viewpoints. This will ultimately result in more sustainable outcomes.

Besides, environmental communication motivates people to take action. Environmental communicators can inspire the public to make positive changes in their daily lives by presenting environmental champions, success stories, and useful advice for sustainable living. By encouraging people to support eco-friendly products, reduce waste, conserve energy, and take part in community clean-up events, environmental communication can help people realise that their actions matter and can help create a healthier planet for future generations.

1.4.6 Encourage the Sharing of Knowledge

Scholars argue that environmental communication brings urgent environmental issues to the public's attention [11]. Therefore, people are made aware of the urgency of taking action to protect our planet by emphasising the effects of environmental degradation and the value of sustainable practices. Environmental communication uses a variety of platforms, including social media, news sources, documentaries, and educational campaigns, to raise awareness of important environmental issues like pollution, deforestation, climate change, and biodiversity loss. Environmental communication motivates people to learn more about these issues and fosters a sense of urgency, which acts as a catalyst for knowledge sharing.

In addition, interaction and conversation between people and communities are fostered by environmental communication. Therefore, individuals are able to exchange ideas, experiences, and knowledge regarding environmental issues by means of discussion, debate, and collaboration platforms. Environmental communication encourages people to come together to address shared environmental challenges by facilitating information exchange through community events, workshops, webinars, and social media campaigns. Environmental communication generates a conducive atmosphere for information exchange and group action by encouraging discussion and active engagement. It also encourages learning about sustainable practices and education. People are empowered to make informed decisions and take responsible actions to protect the environment when they have access to information, resources, and tools for sustainable living. Environmental communication raises people's awareness of environmental issues and gives them the information and tools they need to adopt sustainable behaviours through workshops, online resources, educational programmes, and campaigns. Environmental communication contributes to the development of an informed and environmentally conscious society by supporting education and life-long learning.

1.4.7 Encourage Capacity-Building

The promotion of environmental awareness and knowledge is facilitated by environmental communication [6]. The public can learn more about the condition of the environment and the effects of human activity on the planet by receiving information through a variety of channels, including social media, traditional media, educational initiatives, and community outreach. The public are better able to comprehend the value of environmental preservation and the necessity of taking action to safeguard the environment as a result of this awareness. Cooperation and coordination between various stakeholders are fostered by environmental communication. Indeed, environmental communication plays a crucial role in fostering partnerships and generating collective actions towards sustainable development by serving as a forum for dialogue and discussion among governments, businesses, non-governmental organisations, and community members. Stakeholders can collaborate in implementing practical solutions to environmental challenges and building a more resilient and sustainable future by pooling their knowledge and resources [6].

Also, environmental communication gives communities and individuals the tools they need to take initiative and improve the environment. By providing tools, resources, and information on sustainable practices and green initiatives, environmental communication inspires people to adopt environmentally friendly behaviours and make informed choices in their everyday lives. People who feel empowered have a greater sense of ownership and responsibility for the environment, which inspires them to take up environmental stewardship and push for reform in their local communities.

1.4.8 Increasing Awareness

Public awareness of significant environmental issues and challenges is aided by environmental communication [3]. People are exposed to information about issues like pollution, deforestation, climate change, and endangered species through a variety of communication channels, including news stories, documentaries, social media posts, and educational campaigns. The public are better equipped to protect the environment by taking action and making educated decisions thanks to this knowledge. Individuals can be motivated and inspired to take action for positive environmental change through environmental communication. Communication can inspire people to participate in environmental conservation efforts by telling tales of successful environmental initiatives or emphasising the effects of individual actions on the environment. Through advocacy for environmental policies, support for sustainable businesses, or participation in beach clean-ups, communication can inspire people to make a difference in their communities and beyond.

Besides, fostering a sense of community and shared responsibility for the environment is achieved through environmental communication. By creating platforms for dialogue and discussion about environmental issues, communication can bring people together to share ideas, resources, and solutions. Members of the public can work together to achieve shared environmental objectives and develop a feeling of purpose in safeguarding the planet for coming generations by cooperating and working together.

1.4.9 Conclusion

We have demonstrated that environmental communication is essential for raising awareness and motivating action to tackle environmental issues. This chapter has elucidated the concept of environmental communication and its historical context. Factors that influence the development of environmental communication such as globalisation, industrialisation, and climate change have also explored, highlighting the need for organisations to consider cultural, social, and political contexts when crafting their messages. The chapter has delineated the principles of environmental communication that are efficacious in captivating audiences and encouraging them to undertake action. By understanding these principles, communication organisations can successfully navigate the complexities of discussing environmental issues and drive positive change. Essentially, the chapter has emphasised the significant impact of environmental communication in inspiring individuals and communities to actively engage in actions that support environmental sustainability. This chapter advocates for the utilisation of communication as a strategy to address pressing environmental concerns and inspire cooperative endeavours.

Questions

- What is environmental communication?
- What is the history of environmental communication?
- Mention at least four factors that led to the emergency of environmental communication.
- What are the principles of effective environmental communication?
- Why is environmental communication important?

Recap

Environmental communication is a field focused on how information about the environment is shared, understood, and acted upon. It involves the exchange of ideas among individuals, communities, organisations, and governments concerning environmental issues. The goal is to enhance awareness and encourage action towards sustainability. Environmental communication has become increasingly significant in recent years due to growing concerns about climate change, pollution, and biodiversity loss.

References

1. Antonopoulos N, Karyotakis MA. Data journalism. The SAGE international encyclopaedia of mass media and society. SAGE Publications; 2020. p. 440–1.
2. Flor AG. Environmental communication: principles, approaches and strategies. UP Open University; 2004. https://www.academia.edu/181519
3. Boykoff M. Creative (climate) communications. Cambridge University Press; 2019.
4. Harris US. Engaging communities in environmental communication. Pac J Rev. 2017;23(1):65–79. https://doi.org/10.24135/pjr.v23i1.211.
5. Katz-Kimchi M, Goodwin B. Organising and integrating knowledge about environmental communication. Environ Commun. 2015;9(3):367–9. https://doi.org/10.1080/17524032.2015.1042985.
6. Flor A, Gomez ED, editors. Environmental communication: considerations in curriculum and delivery systems development. University of the Philippines Los Banos - Institute of Development Communication; 1993. https://www.academia.edu/80733434/Environmental_Communication_Considerations_In_Curriculum_and_Delivery_Systems_Development
7. Cox JR. Environmental communication pedagogy and practice. Environ Ed Res. 2018;24(8):1224–7. https://doi.org/10.1080/13504622.2018.1434870.

8. Jacques PJ. Environmental skepticism: ecology, power and public life. Routledge; 2016.
9. Antonopoulos N, Karyotakis MA, Kiourexidou M, Veglis A. Media web-sites environmental communication: operational practices and news coverage. World Media. 2019;2:47–65. https://doi.org/10.30547/worldofmedia.2.2019.3.
10. Murugesan S. Harnessing green IT: principles and practices. IT Prof. 2008;10(1):24–33.
11. Abbati M, Abbati M. The environmental communication under the magnifying lens. Communicating the environment to save the planet: a journey into eco-communication; 2019. p. 3–29. https://doi.org/10.1007/978-3-319-76017-9_1.
12. Lindenfeld LA, Hall DM, McGreavy B, Silka L, Hart D. Creating a place for environmental communication research in sustainability science. Environ Commun. 2012;6(1):23–43. https://doi.org/10.1080/17524032.2011.640702.
13. Pedelty M. Environmental communication and the public sphere. Environ Commun. 2015;9(1):139–42. https://doi.org/10.1080/17524032.2014.1003440.
14. Van Eck CW, Mulder BC, van der Linden S. Echo chamber effects in the climate change blogosphere. Environ Commun. 2021;15(2):145–52.
15. Bognar J, Skogstad G, Mondou M. Media coverage and public policy: reinforcing and undermining media images and advanced biofuels policies in Canada and the United States. Environ Commun. 2020;14(8):1127–44.
16. Lakoff G. Why it matters how we frame the environment. Environ Commun. 2010;4(1):70–81.
17. Lester L. Three challenges for environmental communication research. Environ Commun. 2015;9(3):392–7.
18. Bayes R, Bolsen T, Druckman JN. A research agenda for climate change communication and public opinion: the role of scientific consensus messaging and beyond. Environ Commun. 2023;17(1):16–34.
19. Wikimedia Common. 2022. Industrialisation. https://commons.wikimedia.org/wiki/File:Industrialisation.jpg. Accessed 6 June 2025.
20. The South Durban Community Environmental Alliance (SDCEA). 2024. Community News. https://sdcea.co.za/publication/. Accessed 6 June 2025.

CHAPTER 2

Environmental Challenges in Africa

Introduction
From the dense rainforests of the Congo Basin to the vast savannahs of the Serengeti, Africa boasts a rich diversity of species and ecosystems. This continent is home to remarkable wildlife, including elephants, lions, and numerous bird species. In addition, Africa's unique ecosystems, such as wetlands, grasslands, and mountain ranges, support various life forms that contribute to the planet's biodiversity. However, Africa faces significant environmental problems that threaten not only its ecosystems but also the well-being of its people. This chapter focuses on environmental problems Africa is facing that need to be resolved quickly and collectively. The chapter starts by discussing the concept of the environment, and environmental challenges including deforestation, land degradation, urbanisation, water scarcity, and other factors that pose the biggest threats to Africa's natural systems and human well-being.

2.1 The Concept of Environment

The word "environment" has become a catchphrase in today's society due to the increased awareness of the need to protect and sustain the world. The English term "environment" comes from the French word "environ," which means "surroundings" or "encircling" [1]. This root phrase provides a vivid picture of everything that surrounds people and shapes their life.

The environment is associated with the natural world, which consists of the planet's trees, rivers, mountains, and animals. However, the environment consists of more than just the natural world. It also includes the built environment, which is made up of the infrastructure that people have created, such as

© The Author(s), under exclusive license to Springer Nature Switzerland AG 2025
G. Mutinta, *An African Perspective on Environmental Communication*, https://doi.org/10.1007/978-3-032-05052-6_2

airports, schools, factories, roads, subways, parking spaces, and housing units [2]. Therefore, environment is a complex web of interconnected systems that have varying effects on one another. The origin of the word "environment" is a useful reminder that people do not exist separate from their surroundings but rather are intrinsically tied to the surroundings. There are other factors than the products people consume and the litter they produce which affect the environment. In addition, it is essential that people adapt to the environment and develop and evolve in a sustainable way.

Besides, environmental protection is not only a noble activity but also necessary for the survival of the planet and future generations [3]. Climate change, deforestation, pollution, and biodiversity loss are pressing issues that require immediate attention. People cannot keep abusing the environment and causing it to deteriorate without facing consequences.

All people have a responsibility to support environmental conservation. Whether people are recycling, reducing carbon emissions, conserving water, or encouraging environmentally friendly behaviour, every step counts. To guarantee that everyone has a sustainable future, people must prioritise the state of the planet and work together.

Furthermore, environmental challenges are pressing issues that endanger the well-being and health of our planet and its inhabitants. These issues are typically caused by human and natural activities and might have negative consequences if they are disregarded.

In addition, global concerns that affect individuals beyond national boundaries include pollution, deforestation, and climate change. Because of the interconnectedness of the world, developments in one region can have a profound effect on other regions, requiring international collaboration in the quest for solutions.

Another crucial factor is the complexity of environmental issues. It takes a deep understanding of politics, economics, science, and social dynamics to address these often-difficult concerns. For instance, rising global temperatures are caused by a complicated web of factors such as greenhouse gas emissions, deforestation, and changes in land use. This is called "climate change" [4].

Another feature of environmental challenges is their urgency. Pollution, deforestation, and climate change are already having devastating effects on ecosystems, wildlife, and human cultures worldwide. If action is not taken swiftly, these issues will only worsen and lead to food and water shortages, public health concerns, and increasingly severe natural disasters. The fact that environmental challenges are interconnected, they are often intricately linked to one another. Deforestation, for example, raises the risk of severe and frequent wildfires and droughts, which worsen climate change by releasing carbon into the atmosphere. Owing to this interdependence, solving environmental issues requires a comprehensive approach that takes into account the interconnections of several factors and attempts to address the root causes of issues rather than just their symptoms.

2.2 Environmental Challenges

Environmental challenges in Africa are complex and must be tackled together. From deforestation to climate change, the continent faces a plethora of challenges that endanger the health and well-being of its people and unique species.

2.2.1 *Deforestation*

African forests are home to a wide variety of animal and plant species, many of which are exclusive to the continent. However, in the recent past, though, African forests have been threatened by urban settlements and town expansion, clearing of forests for agricultural purposes, wild forest fires, and massive industrial constructions [5]. All these human activities destroy forests. Now deforestation disrupts ecosystems and destroys habitats, leading to the loss of biodiversity. The delicate balance of nature can suffer greatly when particular species are driven to extinction or are in danger of being extinct due to habitat loss.

Forests help to regulate Earth's climate by taking up carbon dioxide from the atmosphere [6]. When trees are cut down, this carbon is released back into the atmosphere, contributing to global warming and climate change. Deforestation in Africa is making the effects of climate change worse by increasing the frequency and intensity of droughts, floods, and other extreme weather events.

Trees are crucial for preventing soil erosion because of their protective canopy, which aids in moisture retention, and their capacity to anchor the soil with their roots. When trees are destroyed, the soil becomes exposed to the elements, which causes erosion. Soil erosion in some areas can lead to desertification, decreased agricultural yields, and a decline in the fertility of the land.

Besides, forests are essential to the maintenance of the water cycle because they have the power to regulate river and stream flow as well as replenish groundwater reserves. Deforestation reduces the quantity and quality of water in many African locations, impeding this natural process. This has a big effect on drinking water supplies, agriculture, and the overall health of ecosystems.

In addition, for many indigenous people in Africa, forests are vital to their way of life in addition to being a source of food, medicine, and cultural artefacts. Deforestation limits these people's access to forest resources, endangering their traditional way of life. The forced adaptation of indigenous peoples to a rapidly changing environment can lead to social unrest, impoverishment, and forcible migration.

2.2.2 *Land Degradation*

Land degradation is one of the major environmental problems affecting Africa today. It is defined as the decrease in land productivity and environmental quality caused by human activities such as agriculture, urbanisation, and deforestation [6]. This may result in soil erosion, desertification, biodiversity loss, and poor water quality. In Africa, deforestation is a major issue linked to land degradation. Urbanisation, agriculture, and logging have all resulted in the destruction of forests, which has drastically decreased biodiversity and ecological services. Deforestation exacerbates the environmental issues the continent is now facing by contributing to soil erosion, water scarcity, and climate change.

Besides, soil erosion is a major problem in Africa that leads to reduced agricultural productivity and the loss of valuable topsoil. Soil erosion-related land degradation is often caused by inefficient land management practices, such as overgrazing, deforestation, and improper irrigation techniques [6]. Soil erosion not only affects food security but also decreases biodiversity and ecological services in the region (Fig. 2.1).

Land degradation in Africa is also made worse by pollution, which can come in the form of pollutants like industrial waste and agricultural runoff. Pollution in soil and water makes it more difficult for plants and animals to survive. Moreover, communities living near contaminated soil have detrimental health effects because of pollution, highlighting the environmental challenges associated with land degradation in Africa.

Land degradation in Africa has resulted in a decline in biodiversity because of habitat destruction or change caused by human activity [7]. The loss of biodiversity affects ecosystems and increases the danger of extinction for several plant and animal species. Reducing biodiversity increases ecosystems' vulnerability to changes in the environment, making it more difficult to handle issues like food security and climate change [8].

Fig. 2.1 Land degradation. Courtesy of Wikimedia Commons (2025)

2.2.3 *Urbanisation*

Urban regions are recognised for having high pollution levels, which can have severe effects on soil quality [9]. One of the main causes of land degradation is the pollution of the soil, water, and air, as well as improper waste management and vehicle emissions. Contaminants including plastics, pesticides, and heavy metals not only render land unfit for cultivation but also pose a major threat to human and wildlife health.

As cities expand, forests are cleared to make way for additional structures, roads, and other infrastructure. Deforestation reduces biodiversity, tampers with the water cycle, and encourages desertification and soil erosion. Reduced tree cover increases the land's exposure to the elements and makes it more vulnerable to erosion from wind and flood. As a result, the fertility and ability of the soil to support plant life are impaired (Fig. 2.2).

Urbanisation is also putting pressure on natural resources including water, minerals, and energy, leading to overuse of these resources. A lot of African communities have water resources that are depleted or tainted due to excessive extraction for domestic and commercial use [10]. Likewise, mining activities associated with urbanisation can pollute, damage ecosystems, and erode soil. The unsustainable exploitation of resources poses a challenge to the long-term survival of urban areas and has adverse effects on the environment. Urbanisation results in the conversion of agricultural land into urban areas, which causes the loss of fertile soil. Urbanisation is encroaching on arable land, reducing the quantity of land suitable for food production. This disrupts natural habitats and ecosystems, which not only ruins soil but also jeopardises food supplies. Converting land for urban use increases the risk of floods because impermeable

Fig. 2.2 Urbanisation. Courtesy of Freepik (2025)

surfaces, such as concrete, create erosion and sedimentation and prevent rainwater from entering the soil.

2.2.4 *Water Scarcity*

Scholars explain that rising temperatures due to climate change cause water sources such as rivers, lakes, and reservoirs to evaporate more quickly [11]. This puts further strain on the already limited water supplies in Africa's arid and semi-arid regions by lowering the quantity of water available for domestic, industrial, and agricultural use. As a result, farmers are unable to provide their crops with enough water, which reduces agricultural output and raises food insecurity (Fig. 2.3).

The changing weather patterns brought on by climate change, such as erratic rainfall and extended droughts, exacerbate Africa's water scarcity. In many places, the usual rainy season has shortened and become less predictable, leaving farmers without assurances about the availability of water for their crops [12]. Since agriculture is the primary source of income in rural regions, this leads to food shortages, crop failures, and an increase in poverty.

Moreover, deforestation and land degradation—both of which are often exacerbated by climate change—are the causes of Africa's water scarcity. Trees absorb and release water into the atmosphere through a process known as

Fig. 2.3 Water scarcity. Courtesy of Wikimedia Commons (2025)

transpiration, which aids in the regulation of the water cycle. However, widespread deforestation and soil degradation lower trees' water-holding capacity, which lowers the volume of water in rivers and underground aquifers.

In addition, climate change-related changes in weather patterns and rising temperatures have the potential to damage Africa's freshwater supplies and spread waterborne infections. This puts communities who rely on tainted water for drinking and sanitation in grave peril and exacerbates the health risks already linked with water scarcity.

2.2.5 Water Pollution

Water pollution is currently one of Africa's largest environmental issues [12]. The rapid industrialisation and population growth have created a bigger demand than ever for clean water. Nevertheless, due to pollution from a range of causes, including industrial waste, agricultural runoff, and inadequate sewage treatment facilities, the quality of the water in many African countries has worsened [12]. This poses a major threat to aquatic life, human health, and the environment. For example, contaminated water supplies have the potential to spread diseases including cholera, typhoid, and dysentery. Many African communities lack access to clean drinking water, and drinking tainted water can have negative health implications, particularly for young people [13]. In essence, water pollution in Africa poses a threat to aquatic environments. Pollutants including industrial chemicals, herbicides, and heavy metals can accumulate in water bodies and damage aquatic flora and animals. The delicate balance of the environment can be disrupted as a result, leading to a decline in fish populations and other aquatic animals. The loss of biodiversity in water bodies can have a substantial influence on the food chain and ecological stability, highlighting the need for sustainable water management practices (Fig. 2.4).

Thus, contaminated water has a major effect on agriculture. Many farms utilise groundwater sources such as lakes and rivers for irrigation and animal watering. However, if pesticides and other pollutants are introduced into water sources, agricultural production and soil fertility can decrease. Pollutants in agricultural runoff have the potential to affect groundwater sources, exacerbating the problem of water pollution.

Finally, water pollution in Africa has negative economic ramifications [14]. Many communities depend on bodies of water for fishing, transportation, and leisure. Nonetheless, the pollution of these water sources can limit economic opportunities and hinder sustainable development. For example, the decline in fish stocks due to water pollution can have an impact on the livelihoods of fishing communities and the lowering of food security. In addition, polluted rivers can deter tourists and investors, which might lead to a drop in prospective profits and economic expansion.

Fig. 2.4 Water pollution. Courtesy of SDCEA (2024)

2.2.6 *Climate Change*

Africa is experiencing more frequent and severe droughts, which is one of the most evident consequences of climate change [14]. Droughts can have a severe negative impact on agriculture, water resources, and food security. In regions like the Sahel, the Horn of Africa, and Sub-Sahara Africa droughts have resulted in widespread famine, animal deaths, and crop failures, pushing people into poverty and creating social instability [15].

As a result of climate change, many parts of Africa are experiencing increasingly frequent and severe floods. Infrastructural destruction, population displacement, and polluted water supplies have resulted from flash floods, river overflows, and coastal erosion brought on by rising sea levels and strong rainfall [16]. Flooding increases the risk of waterborne diseases like cholera and malaria, which can exacerbate the already unstable situation of medical institutions (Fig. 2.5).

Africa is home to a vast array of flora and fauna, many of which are threatened with extinction as a result of climate change. Owing to ecological disturbance brought on by warming temperatures, deforestation, and habitat loss, animals are being forced to relocate in order to survive [15]. Consequently, there is a decrease in biodiversity, which might potentially impact several aspects of the ecosystem such as soil fertility, pest control, and pollination.

Fig. 2.5 Climate change. Courtesy of Wikimedia Commons (2025)

Besides, climate change is the cause of desertification or the expansion of deserts across Africa [17]. Drought, deforestation, overgrazing, and poor land management practices are turning fertile land into wastelands, making it more difficult for people to raise food and raise animals. Because individuals are forced to travel in search of better opportunities, desertification exacerbates poverty, conflict, and migration. Over the years, Africa's water cycle has changed in numerous ways as a result of climate change, leading to a lack of water [18]. Rising evaporation, erratic rainfall, and glacier melting are making freshwater for drinking, farming, and sanitation less available. This puts additional pressure on the already scarce water supplies, exacerbates conflicts over water rights, and obstructs the development of the economy and public welfare.

2.2.7 Biodiversity Loss

The loss of biodiversity is a primary cause in the decline of significant ecosystems in Africa, such as wetlands, grasslands, and forests [18]. These ecosystems carry out crucial tasks including regulating temperature, purifying water, and stopping soil erosion. The loss of these vital services brought about by a species' extinction upsets the delicate balance of these ecosystems and endangers the continent's ecology and population. As a consequence, Africa's declining biodiversity poses a challenge to livelihoods and food security. Many African societies rely on biodiversity for food, medicine, and other essentials [19]. Communities find it more difficult to meet their basic needs when traditional knowledge on how to use these resources is lost along with species. This can exacerbate the cycle of environmental degradation and biodiversity loss by contributing to poverty, hunger, and malnourishment.

In addition, the dwindling biodiversity of the continent poses a threat to Africa's unique fauna and natural heritage. Africa is home to some of the most famous creatures on the planet, such as elephants, rhinos, and lions. Unfortunately, illegal poaching, habitat degradation, and climate change are putting many species at risk of extinction. The loss of these species will not only have a significant detrimental impact on the ecosystems of the continent but also deprive future generations of Africa's rich biodiversity.

Research suggests that the deterioration of Africa's biodiversity is exacerbating the effects of global warming [20]. Restoring healthy ecosystems is crucial to storing carbon dioxide and mitigating the consequences of global warming. A decline in biodiversity weakens these ecosystems' ability to absorb carbon dioxide, which increases greenhouse gas emissions and quickens global warming. This has a negative impact on Africa's environment, causing more frequent and severe droughts, floods, and other extreme weather events.

Besides, Africa's declining biodiversity is undermining efforts to achieve sustainable development. In addition to the environment, society and the economy are also impacted by biodiversity loss. For example, the loss of pollinators such as bees and butterflies can result in lower agricultural output, which would affect the availability of food and the livelihoods of farmers. The loss of genetic diversity in crops and livestock puts agricultural systems at risk for diseases and pests, threatening both food security and economic stability.

2.2.8 Habitat Destruction

One of the primary ways that habitat degradation affects Africa is through deforestation. The continent Africa is losing its forests to the development of its timber industry, infrastructure, and agricultural sectors [20]. This is happening at a very alarming rate. Numerous plant and animal species—some of which are native to the continent—have therefore lost important habitats. Deforestation contributes to climate change by increasing the frequency of droughts and floods in the region and releasing carbon dioxide into the atmosphere and disrupting the water cycle.

In Africa, mining activity also has a major impact on habitat destruction. Large-scale mining operations usually include the clearing of vast tracts of land, which destroys wetlands, forests, and other important ecosystems. This disturbs wildlife and contaminates soil and water supplies with dangerous chemicals, which is bad for human and animal health. What is more, soil erosion and deforestation brought on by mining operations worsen the region's habitat loss.

Besides, urbanisation is a significant factor leading to habitat deterioration in Africa [21]. With the rapid population growth across the continent, cities and towns are encroaching on vital wildlife habitats, turning formerly uninhabited wilderness areas into urban settlements. Conflicts between people and wildlife are made worse by the necessity for animals to battle for resources and navigate strange and occasionally dangerous urban settings. Urbanisation hinders species' ability to move and reproduce in addition to causing habitat fragmentation and pollution, which ultimately threatens their existence.

Further, commercial farming and livestock grazing are two additional prominent agricultural activities that greatly contribute to environmental degradation in Africa [22]. Large-scale agriculture usually involves clearing forests and grasslands to create place for crops and pastures, which lowers biodiversity and eliminates important habitats for a range of species. Chemical fertilisers and pesticides used in agriculture have the potential to contaminate soil and water supplies, aggravating environmental deterioration and posing health risks to both humans and animals.

Another cause of the destruction of African ecosystems is the building of power plants, dams, and roads [23]. These projects usually require clearing land and rerouting rivers, disturbing ecosystems and uprooting species in the process. Infrastructure development also divides ecosystems, isolating populations of species and making it harder for them to find food, mates, and suitable housing. As a result, there might be a decline in biodiversity and a greater chance of many local species becoming extinct.

2.2.9 Air Pollution

Air pollution is becoming an increasingly serious environmental issue in Africa, posing a threat to public health and the ecology [24]. Air pollution levels are growing across the continent as a result of rapid industry, urbanisation, energy waste, and inadequate waste management.

Air pollution is a major health risk in Africa. According to the World Health Organisation, millions of avoidable deaths worldwide are attributed to air pollution each year, with a significant number of these deaths occurring in Africa [25]. When released into the atmosphere, pollutants such as particulate matter, sulphur dioxide, and nitrogen oxides have the potential to cause respiratory disorders, cancer, and heart disease. The populations most vulnerable to the harmful effects of air pollution include the young, the old, and those with underlying medical conditions (Fig. 2.6).

Besides, air pollution affects both human health and the ecosystem [24]. Pollutants can contaminate plant, soil, and water sources when they are released into the atmosphere, which reduces biodiversity and harms the health of ecosystems. This might have negative effects on wildlife populations, agricultural productivity, and the overall ecological balance. Increased greenhouse gas emissions and worsening global warming are two possible outcomes of high air pollution, and both can contribute to climate change.

African cities are growing increasingly developed and have a propensity to absorb and hold heat, raising temperatures and contributing to air pollution. The cause of this phenomenon, referred to as "urban heat islands," is air pollution [26]. As individuals and organisations try to stay cool by air conditioning and other methods, this can result in illnesses linked to heat, worsened air quality, and more energy use. Urban heat islands are particularly dangerous in

Fig. 2.6 Air pollution. Courtesy of Freepik (2025)

densely populated regions lacking in green space because they exacerbate the negative effects of air pollution on the environment and human health.

In addition, air pollution is a social justice issue in Africa [27]. Low-income communities and marginalised individuals are disproportionately affected by air pollution due to factors including living near industrial sites, lacking access to healthcare, and having inadequate housing. These groups usually bear the brunt of environmental injustices, with higher rates of pollution-related illnesses and poorer living conditions.

2.2.10 Natural Disasters

In Africa, natural disasters pose a serious threat to the ecology, economy, biodiversity, and general well-being of the continent [20]. Apart from disrupting ecosystems, natural catastrophes such as hurricanes, earthquakes, floods, and droughts also have a big impact on human populations. Natural disasters exacerbate food insecurity in the continent. Severe weather conditions like droughts and floods can destroy crops, disrupt food supplies, and result in mass malnutrition and hunger. In a continent where millions of people primarily depend on agriculture for both their livelihood and income, any disruption to food supply can have catastrophic consequences.

Natural disasters can lead to biodiversity loss and ecosystem deterioration. Hurricanes and wildfires have the potential to destroy important ecosystems, which would result in the loss of plant and animal species in Africa, a continent abundant in wildlife and unspoiled scenery [7]. This loss of biodiversity can eventually have a severe effect on the ecological balance and the continent's ability to withstand natural disasters.

Additionally, natural disasters can have a big effect on Africa's water supply [26]. Droughts can cause shortages of water, while floods can contaminate water supplies with pollutants and debris. A major risk to public health might arise from the contamination of clean, safe drinking water sources. The destruction of water infrastructure during storms can make the water crisis in many African countries worse. Indeed, natural disasters can have extremely detrimental consequences on the economy of Africa. These billion-dollar damages can result from the demolition of homes, businesses, and infrastructure, seriously harming regional economies and obstructing growth plans. In countries with limited resources and insufficient protocols for disaster preparedness, the expense of repair and rebuilding might be prohibitive, exacerbating poverty and inequality.

Natural disasters can also be blamed for climate change and environmental degradation in Africa [27]. The frequency and intensity of extreme weather events are increasing, and these catastrophes are often linked to unsustainable resource usage and global warming. Consequently, the continent faces an endless cycle of environmental issues that both precipitate and exacerbate natural disasters.

2.2.11 Energy Accessibility

Many African countries largely rely on fossil fuels like coal, oil, and natural gas to meet their energy needs [13]. Fossil fuel combustion releases greenhouse gases, such as carbon dioxide, into the atmosphere, which in turn causes climate change and global warming. This affects not just the local surroundings but also the planet at large.

In many parts of the continent, wood and charcoal are the primary energy sources used for heating and cooking [11]. Widespread deforestation results from the need to supply fuel with more trees being cut down. Deforestation destroys significant ecosystems and habitats and decreases the planet's ability to absorb carbon dioxide from the atmosphere, which exacerbates climate change.

Besides, millions of Africans depend on traditional biomass fuels like wood, charcoal, and dung for cooking and heating, which pollutes indoor air and has a negative impact on their health [15]. These fuels emit smoke that is laden with volatile organic compounds, particular matter, and carbon monoxide—all of which are harmful pollutants. These toxins can cause respiratory conditions, heart problems, and even premature death.

Scholars explain that fossil fuels are widely exploited and used throughout Africa in order to provide energy [15]. This can result in runoff, spills, and leaks that pollute water sources. For instance, oil spills harm fish and other marine life and contaminate water sources, destroying aquatic ecosystems. Further degradation of water quality can come from mining activities linked to energy production, which release potentially dangerous materials into rivers and lakes.

Besides, many people in Africa lack access to sustainable energy sources including hydropower, solar electricity, and wind power, despite the continent's abundance of renewable energy resources [19]. This prolongs the transition to a low-carbon economy and keeps us dependent on unclean and inefficient energy sources. By investing in energy efficiency and renewable energy technologies, African countries can improve the well-being of their people and reduce their environmental footprint.

2.2.12 *Desertification*

Research shows that desertification harms Africa through biodiversity loss [12]. With the loss of vegetation cover and a deterioration in soil quality, many plant and animal species are losing their habitat. This leads to the extinction of species that are vital to the ecosystem and disturbs the delicate balance of natural processes that sustain life in these places.

Another consequence of desertification in Africa is population displacement. Communities are forced to move in search of better opportunities as water sources dry up and land becomes less fertile. This strains already crowded urban regions and upsets cultural norms and conventional ways of life, which in turn leads to social unrest and violence (Fig. 2.7).

Desertification also has a major influence on food security in Africa [9]. A decrease in agricultural output brought on by less productive arable land and limited water supply leads to food shortages and famine. This is particularly

Fig. 2.7 Desertification. Courtesy of Wikimedia Commons (2025)

dangerous in a region where farming provides a large share of the population's income and where food access is already limited.

Beyond these direct repercussions, desertification in Africa has broader implications for global climate change. Loss of plant cover and declining soil quality result in the release of carbon dioxide into the atmosphere, aggravating climate change and contributing to the greenhouse effect. This affects not only Africa's climate patterns but also global weather systems and sea levels.

Finally, human rights are relevant to the problem of desertification in Africa. As a result of the lack of resources and support necessary to cope with the changing environment, the most vulnerable and disadvantaged groups are often those that suffer the most from environmental difficulties. This perpetuates cycles of poverty and inequality while also undermining people's rights to live in a clean and healthy environment.

2.2.13 *Wildlife Trafficking*

Numerous species are endangered, and the delicate ecological balance is seriously threatened by the illegal wildlife trade in Africa [16]. The illegal wildlife commodities trade poses a major danger to both the animals and the biosphere as a whole. To take an example, biodiversity is reduced as a result of animal trafficking. Africa is home to a wide variety of species, many of which are already endangered or in risk of going extinct [17]. The illegal wildlife item trade exacerbates this problem by further reducing the numbers of these vulnerable animals. This disturbs the delicate ecological balance in many locations, which can lead to the extinction of species, and therefore can have a significant impact on the ecosystem (Fig. 2.8).

Fig. 2.8 Wildlife trafficking. Courtesy of Wikimedia Commons (2025)

Animal trafficking contributes to the spread of disease. People can get zoonotic infections from a variety of wildlife species by coming into contact with sick animals or animal products. Animals are often kept in cramped, dirty conditions; thus, the illegal wildlife commodities trade creates an ideal environment for the development of these diseases. This poses a significant risk to the general public's health and might lead to the spread of deadly diseases, which could have disastrous consequences for both humans and wildlife.

Studies indicate that wildlife trafficking disrupts ecosystems and habitats [17]. Many species are necessary for the proper operation of their specific ecosystems because, among other things, they disperse seeds, pollinate plants, and control insect populations. Due to illegal trafficking, these species can be removed from their natural habitats, upsetting the ecosystem. This might endanger other species whose survival depends on these mutually reinforcing relationships, perhaps leading to the loss of significant ecological services.

Besides, wildlife trafficking provides funding for corruption and organised crime. Well-oiled criminal networks usually participate in the illegal trade in wildlife items by taking advantage of weaker species. When these criminal networks bribe off officials and skirt the law to smuggle animal products across borders, they often escape punishment. The campaign against wildlife trafficking is hampered by corruption, which also perpetuates an unstoppable cycle of environmental damage.

Lastly on this score, the wildlife trafficking deprives neighbouring communities of vital resources [12]. For many rural people in Africa, animals are a source of food, income, and cultural activities. The illegal wildlife commodities trade not only puts numerous species in jeopardy but also deprives local communities of resources needed to maintain their way of life. This might make poverty, food insecurity, and civil unrest worse and worsen Africa's environmental issues.

2.2.14 Plastic Pollution

One of the most noticeable consequences of plastic pollution in Africa is the obstruction of rivers and oceans. As a result of plastic bags, bottles, and other trash clutter rivers, lakes, and oceans, marine life finds it impossible to survive. Seabirds, dolphins, and sea turtles sometimes mistake plastic for food, which can lead to ingestion and eventual death. The loss of marine biodiversity has severe consequences for the fragile ecosystems that depend on healthy seas [3].

The quantity of plastic waste that ends up in the oceans and streams can have a major negative influence on Africa's food security. Particles of plastic can make their way up the food chain and end up in people's diets through fish and other marine life. Consuming plastic-contaminated seafood can have detrimental effects on one's health, including hormone abnormalities and other long-term conditions [6]. This poses a danger to the welfare and means of sustenance of coastal populations who depend on seafood.

In addition, greenhouse gas emissions, a consequence of the production and disposal of plastic products, are one of the main contributors to climate change [21]. Many African practices for disposing of plastic waste, such open burning and landfilling, release harmful gases into the atmosphere, exacerbating the climate crisis. As the globe heats, the impacts of plastic pollution—increased floods and coastal erosion, for example—will get worse, further damaging the environment (Fig. 2.9).

Travellers from all over the world are drawn to Africa by its natural beauty and abundance [16]. Plastic pollution has the potential to harm local economies that rely on tourism revenue and to drive away visitors by deteriorating these picturesque locations. In addition to beaches littered with plastic garbage being unsightly, visitors can run the risk of experiencing health issues. Since plastic pollution seriously jeopardises the growth of sustainable tourism in Africa, action must be taken to ensure the industry's continued progress.

Scholars hold that the disproportionate effects of plastic pollution on Africa's most vulnerable communities exacerbate poverty and inequality [8]. To make a meagre income, informal garbage pickers routinely endanger their lives by searching through mountains of plastic waste. Poor waste management techniques cause plastic pollution to build up in low-income communities, putting residents' health and the environment at risk. Plastic pollution must be addressed to promote social justice and provide a sustainable future for all Africans.

2.2.15 Energy Insecurity

Energy instability remains a pressing issue for the African continent, posing significant environmental challenges. Based on recent figures, around 70% of people in sub-Saharan Africa lack access to reliable energy sources [22]. This

Fig. 2.9 Plastic pollution. Courtesy of SDCEA (2024)

energy deficit not only affects people's daily lives but also impedes sustainable development and exacerbates environmental degradation.

One major way that energy instability impacts the environment in Africa is the overuse of traditional biomass fuels, such wood, charcoal, and animal dung, for cooking and heating [7]. As a result, the overuse of these resources results in deforestation, soil erosion, and biodiversity loss, which further exacerbates climate change and environmental degradation.

A shortage of clean and reliable energy sources contributes to the widespread use of dangerous and inefficient technology like kerosene lighting and diesel generators [11]. By releasing harmful chemicals into the environment, these energy sources contribute to respiratory conditions, air pollution, and climate change. The resulting health repercussions not only negatively impact people's well-being but also put further burden on already fragile healthcare systems.

Furthermore, Africa's energy deficit makes it challenging to promote sustainable development and address environmental problems such as water scarcity and food poverty [8]. Without consistent electrical supplies, communities struggle to set up waste management facilities, agricultural irrigation methods, and clean water and sanitation systems. This makes it harder to achieve the goals of sustainable development and worsens the cycle of poverty.

In addition, Africa's lack of energy access impedes the adoption of renewable energy technology, such as wind turbines and solar panels, which have the potential to reduce climate change and enhance environmental sustainability [25]. Due to the high upfront costs and lack of funding, many communities are unable to transition to sustainable energy sources, which worsens environmental issues and increases their reliance on fossil fuels. Energy instability worsens global environmental issues like climate change while also having an immediate impact on African inhabitants. Greenhouse gas emissions are a contributing element to the effects of climate change on individuals who are already vulnerable, as the continent is heavily dependent on fossil fuels for energy production. Improving environmental sustainability and mitigating the consequences of global warming require lowering energy poverty in Africa.

2.3 Conclusion

Africa, a continent with a diverse range of species and abundant natural resources, is facing numerous environmental challenges which threaten its well-being and ecosystems. Deforestation, a major issue primarily caused by agricultural expansion, urbanisation, and logging, destroys wildlife habitats and contributes to climate change. Land degradation, such as soil erosion, desertification, and water source degradation, is another major concern affecting agricultural yields and the livelihoods of a large percentage of the population. Urbanisation, driven by rapid population growth and urbanisation, leads to increased pollution, habitat destruction, and a strain on natural resources. Poor waste management practices in many African cities exacerbate these issues, causing environmental pollution and health risks for residents. Water scarcity is another growing concern, exacerbated by climate change, population growth, and poor water management practices. The lack of access to clean and safe drinking water is a major challenge for the region's residents. Threats from poaching, illegal wildlife trade, and pollution are causing the decline of many species, including elephants, rhinos, and lions, which are iconic symbols of Africa's wildlife heritage. A collaborative effort from governments, organisations, and individuals is needed to address these environmental problems, as the future of Africa's natural systems and human populations depends on swift and cooperative action.

Recap
Africa is a continent rich in biodiversity and natural resources, but it faces a range of environmental challenges that threaten both its people and its wildlife. Understanding the environment involves examining the interactions between living organisms and their physical surroundings. The continent has been grappling with several significant environmental issues, including deforestation, land degradation, urbanisation, and water scarcity. Each of these challenges demands urgent action.

Questions
- What is your understanding of the concept of environment challenges?
- What is being threatened by the multitude of environmental challenges facing Africa?
- What are the main environmental problems Africa is facing that need to be resolved quickly and collectively?

References

1. Types of environmental issues: meaning, pollution, videos, examples. Toppr-guides. 2018.
2. Obi CI. Environmental movements in sub-Saharan Africa: a political ecology of power and conflict. UN Research Institute for Social Development; 2005.
3. Bosetti V, Lebowski R, editors. Deforestation and climate change. Edward Elgar; 2010.
4. Banda Jr, S. Deforestation reaches worrying level—UN. africanews.com. 2018. https://web.archive.org/web/20081008083104/http:/www.africanews.com/site/list_messages/18831.
5. Roe S, Streck C, Beach R, et al. Land-based measures to mitigate climate change: potential and feasibility by country. Glob Chang Biol. 2021;27(23):6025–58.
6. Deforestation statistics for Nigeria. Mongabay. 2021. https://worldrainforests.com/deforestation/archive/Nigeria.htm.

7. Wana F, Senapathy M. Small-scale Irrigation Utilisation by Farmers in Southern Ethiopia: monograph. Primedia eLaunch LLC; 2023. p. 174. https://doi.org/10.36074/ssiubfise-monograph.2022.
8. Oyediji OT, Adenika OA. Forest degradation and deforestation in Nigeria; poverty link. Int J Multidiscip Res Anal. 2022;5(10):2837–44.
9. Hillstrom K, Hillstrom LC. North America: a continental overview of environmental issues. USA: Bloomsbury Publishing; 2003. p. 85–6.
10. Favretto N, Dougill AJ, Stringer LC, Afionis S, Quinn CH. Links between climate change mitigation, adaptation and development in land policy and ecosystem restoration projects: lessons from South Africa. Sustainability. 2018;10(3):779.
11. IFA: International Fertilizer Industry Association. Soil degradation in Africa Low use of fertilizers is one of the main causes of environmental degradation. Fertilizer.org. IFA. 2013. https://web.archive.org/web/20130510070935/http:/www.fertilizer.org/ifa/HomePage/SUSTAINABILITY/Soil-degradation-in-Africa
12. Land and environmental degradation and desertification in Africa. FAO; 2013. https://www.fao.org/4/X5318E/x5318e02.htm.
13. IFA: International Fertilizer Industry Association. Soil degradation in Africa Low use of fertilizers is one of the main causes of environmental degradation. Fertilizer.org. IFA; 2013.
14. Soil degradation. Goodplanet.info; 2009. https://web.archive.org/web/20130518084447/http:/www.goodplanet.info/eng/Pollution/Soils/Soil-degradation
15. Lanly JP. Tropical forest resources food and agriculture organisation of the United Nations. Rome, Italy: United Nations; 1982. http://www.fao.org/docrep/015/an778e/an778e00.pdf
16. Wood TS, Baldwin S. Fuelwood and charcoal use in developing countries. Annu Rev Energy. 1985;10(1):407–29.
17. Jacobson MZ. On the causal link between carbon dioxide and air pollution mortality. Geophys Res Lett. 2008;35(3):S2CID 16440166. https://doi.org/10.1029/2007GL031101.
18. World Health Organisation. Regional burden of disease due to indoor air pollution. 2013. https://web.archive.org/web/20040701221414/http:/www.who.int/indoorair/health_impacts/burden_regional/en/
19. Cukrowska EM, Yalala BN, Tutu H, Chimuka L. Spatial distribution and source identification of mercury in dust affected by gold mining in Johannesburg, South Africa. In: Proceedings of the international conference of recent trends in environmental science and engineering; 2017. p. 104.
20. Lusilao-Makiese JG, Cukrowska EM, Tessier E, Amouroux D, Weiersbye I. The impact of post gold mining on mercury pollution in the West Rand region, Gauteng, South Africa. J Geochem Explor. 2013;134:111–9.
21. Papu-Zamxaka V, Mathee A, Harpham T, et al. Elevated mercury exposure in communities living alongside the Inanda Dam, South Africa. J Environ Monit. 2010;12(2):472–7.
22. Liousse C, Assamoi E, Criqui P, Granier C, Rosset R. Explosive growth in African combustion emissions from 2005 to 2030. Environ Res Lett. 2014;9(3):035003.
23. Guide on ambient air quality legislation – Air pollution series. United Nations Environment Programme; 2023. https://doi.org/10.59117/20.500.11822/42536. ISBN 978-92-807-4029-5.
24. Scorgie Y. Air quality and regulation. National Association for Clean Air; 2013.

25. Kang H, Zhu B, Zhu C, de Leeuw G, Hou X, Gao J. Natural and anthropogenic contributions to long-term variations of SO2, NO2, CO, and AOD over East China. Atmos Res. 2019;215:284–93. https://doi.org/10.1016/j.atmosres.2018.09.012. ISSN 0169-8095
26. Hillstrom K, Hillstrom LC. The Worlds environments. A continental overview of environmental issues. Bloomsbury Publishing; 2003. p. 207–8.
27. Fisher S, Bellinger DC, Cropper ML, et al. Air pollution and development in Africa: impacts on health, the economy, and human capital. Lancet Planet Health. 2021;5(10):e681–8. https://doi.org/10.1016/S2542-5196(21)00201-1.

CHAPTER 3

The Role of the Media in Environmental Communication

Introduction
The media has always played a crucial role in shaping public opinion and raising awareness of important issues. In terms of environmental communication, attitudes, actions, and policies can be influenced by the media, which can have a big impact on how the world is. Traditional media has been a trustworthy source of news and entertainment for a long time. These platforms have the ability to reach a wide audience and have a substantial impact on public opinion. Digital media, on the other hand, can allow for more dynamic and participative information dissemination and audience interaction. With this kind of media, real-time updates, interactive content, and a worldwide audience reach are all achievable. This chapter discusses the role of the media in environmental communication. In the first portion of the chapter, the role of the traditional media in environmental communication is discussed. This is followed by a discussion of the role of digital media in environmental communication. The chapter ends with a discussion of successful environmental communication initiatives and finally a discussion on examples of individual environmental advocates in Africa.

3.1 The Role of the Traditional Media in Environmental Communication

Traditional media has conventionally been a useful tool for shaping public perception and encouraging behavioural change, particularly when it comes to environmental issues. Even in the current digital age, where information is

© The Author(s), under exclusive license to Springer Nature Switzerland AG 2025
G. Mutinta, *An African Perspective on Environmental Communication*,
https://doi.org/10.1007/978-3-032-05052-6_3

readily available at our fingertips, traditional media continue to be essential for communicating environmental issues.

3.1.1 Television

Television media is one of the best resources for educating the public about environmental issues. For example, television media can play a significant role in advancing environmental issues and mobilising public support for conservation and legislative reforms. By employing investigative reporting and in-depth coverage of environmental issues, television shows can highlight the importance of protecting the environment and inspire viewers to get involved in environmental activism [1]. Television media has the potential to inspire viewers to join the movement towards a more sustainable and environmentally friendly future by showcasing the work and success stories of environmental organisations.

Television media serves as a watchdog by making companies, governments, and other economic sectors answerable for their environmental deeds. Through investigative journalism and exposés, television shows can reveal instances of corporate misconduct, criminal enterprises, and environmental harm. Television media can promote more transparency and accountability in environmental decision-making by bringing these issues to the public's notice. This might ultimately lead to more sustainable and responsible actions.

Besides, it is possible to see on television cutting-edge technology and solutions that promote environmental sustainability and resource conservation. Television series can highlight creative ways to deal with environmental problems, such as recycling programmes, strategies to reduce litter, green construction practices, and the use of renewable energy sources. By presenting best practices and success stories, television media can inspire viewers to adopt eco-friendly solutions and support the transition to a more sustainable society.

Further, television media can encourage community involvement and discussion on environmental issues by bringing together a range of voices and points of view to examine strategies and solutions for environmental preservation. Through talk shows, panel discussions, and town hall meetings, television programmes can provide a platform for professionals, activists, politicians, and citizens to express ideas, work together on environmental projects, and trade experiences [2]. By promoting inclusive and open communication, television media can enable communities to work together towards common goals and enhance the environment.

Television has the power to influence public opinion and shape governmental policy in relation to environmental concerns. By highlighting environmental issues and showcasing effective conservation efforts, television media can impact legislative changes that promote sustainable practices and protect the environment.

Moreover, the behaviour of viewers might be influenced by television programmes that promote sustainable behaviours and lifestyles. Through the promotion of eco-friendly goods, events, and initiatives, television media can encourage viewers to take on more ecologically conscious habits and reduce their ecological footprint.

3.1.2 Radio

Radio has always been a powerful tool for communication, and it continues to have a big influence on listener participation and the movement for change [3]. For several reasons, including raising awareness of pressing environmental issues, radio is an effective medium for environmental communication. By playing instructional programmes and news articles, radio stations can inform listeners about the gravity of issues such as pollution, deforestation, climate change, and animal conservation. With its captivating stories and factual reporting, radio can increase public awareness of these problems and motivate listeners to take action.

Education is a key element of environmental communication, and radio can be a very effective medium for educating people about conservation efforts, sustainable development, and environmental science. By presenting experts, environmentalists, and campaigners on their programmes, radio stations can help inform listeners about the importance of protecting the environment and the steps they can take to have a positive impact. A radio programme can also inform listeners about local environmental initiatives and events, as well as resources available to assist neighbourhood activism.

Besides, advocacy is yet another crucial use of radio in environmental communication [4]. Public awareness of important environmental issues can be increased by using radio to magnify the voices of advocacy groups, community organisers, and environmental activists. Stakeholders can explore sustainable practices, legislative reforms, and ways to promote environmental justice in panel discussions, interviews, and radio talk programmes (Fig. 3.1).

Community involvement is essential to effective environmental communication, and radio can be a key medium for building cross-community linkages and motivating collective action. Radio programmes can highlight neighbourhood-based initiatives, tell the stories of local environmental heroes, and provide people a platform to discuss environmental issues. By promoting collaboration and communication, radio can help communities grow strong and resilient and committed to protecting the environment.

Research shows that radio's main function in environmental communication can be to inspire action [5]. Radio stations have the capacity to inspire listeners to take proactive steps towards sustainability, which can lead to real societal change. Whether it is encouraging listeners to reduce their carbon footprint, volunteer for cleaning efforts, or make donations to environmental charities, radio has the capacity to motivate listeners to actively participate in environmental stewardship and enhance their communities.

In addition, as radio fosters optimism and hope for the future, recognising successes is crucial to environmental communication [6]. Radio has the power

Fig. 3.1 Radio image. Artificial Intelligence Generated (2025)

to share uplifting stories of environmental innovation, conservation, and restoration that highlight the incredible achievements made by individuals, organisations, and communities to protect the environment. By showcasing these successes, radio can inspire listeners to continue working towards solutions for environmental issues.

3.1.2.1 The Role of Community Radio Stations in Environmental Communication

Community radio stations play a vital role in environmental communication. By providing accessible forums for raising awareness, amplifying marginalised viewpoints, educating the public, and driving action, these stations have the ability to encourage environmental activism and contribute to a more sustainable future. It is vital to support and empower community radio stations in their efforts to promote environmental stewardship and protect the earth.

Radio Okapi

Radio Okapi is a community radio station located in the Democratic Republic of the Congo (DRC) that has been instrumental in promoting knowledge on environmental conservation and sustainable practices [4]. Radio Okapi has substantially enhanced environmental awareness and advocated for sustainable practices in the DRC via its extensive programming and broadcasts.

Radio Okapi underscores the need of environmental conservation for future generations. The radio station often conducts discussions and interviews with experts on environmental issues, highlighting the impacts of deforestation, pollution, and climate change on the ecosystems of the DRC [4]. Radio Okapi aims to inspire residents to undertake proactive initiatives and implement positive change in their communities by disseminating information on pressing environmental concerns.

Furthermore, Radio Okapi emphasises the need of promoting ecological practices, such as recycling, energy conservation, and effective waste management [4]. The radio station provides pragmatic advice on how individuals may reduce their environmental footprint and promote a more sustainable future via its programming. Radio Okapi is cultivating a culture of environmental stewardship in the DRC by providing its listeners with the knowledge and tools necessary to adopt sustainable habits [4].

- **Lessons**

In addition to its discourse on environmental conservation and sustainable practices, Radio Okapi imparts other valuable lessons as a community radio station. Radio Okapi demonstrates the effectiveness of grassroots media in promoting awareness and initiating positive change within local communities. Through active engagement with its audience and the discussion of relevant issues, the radio station successfully reaches a wide audience and mobilises support for environmental efforts.

Radio Okapi emphasises the need of collaboration and alliances in promoting environmental sustainability. The radio station partners with governmental entities, non-profit organisations, and community groups to amplify its message and implement effective solutions to environmental challenges. By fostering partnerships with key stakeholders, Radio Okapi uses resources and expertise to further its environmental advocacy efforts.

In addition, Radio Okapi exemplifies the need of cultural sensitivity and local context in promoting environmental awareness. The radio station tailors its messaging and programming to correspond with the cultural values and beliefs of its audience, hence increasing relatability and engagement for listeners. Radio Okapi promotes substantial behavioural change and advocates for sustainable practices in the DRC by incorporating local perspectives and traditions into its environmental messages.

Radio Okapi illustrates the impact of storytelling in local languages and narrative in shaping public perceptions and attitudes about the environment [4]. The radio station utilises compelling narratives and personal anecdotes in its programming to convey the significance of environmental issues and inspire listeners to take action. Utilising the storytelling qualities of radio, Radio Okapi successfully engages its audience emotionally and promotes substantial changes in environmental behaviour.

Furthermore, Radio Okapi uses several communication tools, including radio broadcasts, social media, and community events, to convey its environmental concerns to a wider audience. The radio station is improving the effectiveness of its programming and encourages participation in sustainability initiatives by using many channels to provide information and interact with its audience.

Bush Radio Station

Bush Radio in South Africa is a community radio station that broadcasts messages on environmental conservation and sustainable practices [7]. This radio station has played a crucial role in educating the public on the need of environmental stewardship and the implementation of sustainable practices in daily life.

Bush Radio is however facing challenges related to funding [4]. As a community radio station, Bush Radio relies heavily on donations and sponsorships to continue its operations. This has a significant challenge, as funding is often limited and unpredictable. Additionally, Bush Radio is struggling with limited resources and infrastructure, which hinders its ability to reach a wider audience and make a greater impact.

- **Lessons**

The principal lesson gleaned from Bush Radio is the need of community engagement. The radio station has included the local community in its environmental conservation initiatives and campaigns. Bush Radio has effectively increased awareness and educated the people about the importance of environmental conservation and the implementation of sustainable practices via community participation. Involving the community in environmental initiatives cultivates increased awareness and support for conservation endeavours.

Bush Radio has shown the need of using media platforms to convey messages on environmental conservation. The station has successfully engaged a wide audience and increased awareness of environmental issues via its radio programmes and initiatives. This highlights the media's significance in educating the public and inspiring participation in environmental conservation initiatives. Media outlets can be powerful tools for promoting environmental awareness and sustainability efforts [1].

The final lesson from Bush Radio emphasises the need of promoting local efforts for environmental protection. The radio station has consistently supported local environmental projects and campaigns in South Africa. By promoting local initiatives, Bush Radio has cultivated a sense of community ownership and responsibility for environmental conservation. This demonstrates that supporting local initiatives may significantly impact the development of environmental protection and sustainability policy.

Koch FM

Koch FM is a Kenyan community radio station committed to promoting environmental conservation and sustainable practices [4]. The station has shown perseverance and dedication in achieving its goal. Koch FM provides substantial insights into the development of environmental consciousness.

A key challenge for Koch FM is the limited scope of its broadcast [4]. Community radio stations in Kenya have difficulties in acquiring broadcasting licences and face governmental restrictions, impeding their capacity to reach a broader audience. Furthermore, securing money is a continual struggle for the station, since it relies heavily on donations and sponsorships for its viability [4].

- **Lessons**

Despite these challenges, Koch FM has effectively secured a unique place in the Kenyan media landscape. The station transmits many programming focused on environmental conservation and sustainability, including discussions in different languages on climate change, conservation efforts, and eco-friendly living options. Koch FM collaborates with local environmental organisations and experts to provide accurate and up-to-date information on environmental issues to its audience.

Koch FM underscores the need of individual accountability in environmental preservation. Listeners are encouraged to adopt simple practices in their daily lives, such as reducing waste, conserving energy, and supporting sustainable businesses. Koch FM aims to promote grassroots positive change by highlighting the impact people can have on environmental conservation [4].

Further, Koch FM promotes environmental conservation and underscores the need of sustainable ideals throughout all sectors of society. The station advocates for sustainable agriculture, renewable energy, and ethical consumerism. Koch FM highlights the interconnectedness of environmental and social issues, encouraging listeners to reflect on their overall impact on the planet [8].

Koch FM underscores the importance of collaboration and alliances in promoting sustainable practices. By collaborating with local organisations, experts, and businesses, the station can amplify its message and broaden its audience reach.

Besides, Koch FM illustrates the effectiveness of using media as a vehicle for social change. Community radio stations have the ability to educate and inspire listeners on important issues, promoting awareness and stimulating action [3].

Finally, Koch FM illustrates the importance of perseverance under adversity. Despite the challenges it faces, the station continues to advocate for environmental preservation and sustainability, demonstrating a commitment to making a positive impact on the world.

3.1.3 Newspaper

Newspapers have always been an essential tool for communication and information sharing, influencing public opinion and perception [5]. Newspapers still hold a particular role as a means of communicating environmental issues, even in the current digital age where information can be obtained with only one click.

Newspapers can play a major role in public education on environmental issues such as pollution, deforestation, climate change, and animal conservation. Newspapers that publish editorials, opposite the editorial page or op-eds, and stories on these topics can inform readers on how human activity affects the environment.

Studies show that newspapers can also be utilised as a platform to highlight initiatives and endeavours that are effectively enhancing the environment [5]. Newspapers that feature these success stories can inspire readers to get involved and make improvements in their own communities (Fig. 3.2).

Moreover, newspapers can be a useful tool for advocating for and taking action on environmental issues. By publishing articles urging readers to take action or outlining how they can support environmental issues, newspapers can mobilise the public and promote legislative improvements that will enhance the environment. Newspapers can also be a very valuable tool for holding politicians and government agencies accountable for their actions—or lack thereof—on environmental issues. Newspapers that cover government actions, decisions, and policies which affect the environment can help raise public understanding of politics and encourage public engagement in it [9]. In addition, newspapers can promote eco-friendly lifestyle choices and behaviours that

Fig. 3.2 Newspapers. Courtesy of Wikimedia Commons (2025)

reduce the impact of human activity on the environment. Newspapers that publish articles on topics like recycling, energy conservation, sustainable agriculture, and eco-friendly products can educate readers on practical ways to lead a more sustainable life. Research shows that newspapers can provide a platform for discussion and debate on environmental concerns, giving readers a chance to voice their thoughts, bring up relevant topics, and take part in insightful debates [5]. Through promoting an inclusive and transparent discourse, newspapers can aid in the establishment of consensus and the encouragement of collective action on environmental challenges.

3.1.4 Magazine

Reading magazines can teach readers about environmental issues and how they impact the environment. Magazines can provide incisive information through in-depth articles, infographics, and expert interviews on topics like pollution, biodiversity loss, and climate change. Magazines can motivate readers to take action by stressing the importance of these issues and presenting the information in an engaging and approachable way [7].

Magazines have the power to highlight individuals and organisations who are positively impacting the environment and to tell success stories. With their coverage of scientists, environmentalists, and local leaders, magazines can inspire readers to become involved and make a difference in their areas. Some readers can find inspiration and hope in these stories if they are feeling overwhelmed by the magnitude of the environmental problems they face.

Additionally, magazines can serve as a forum for promoting sustainable and eco-friendly products and practices [10]. Magazines that feature articles on green living, eco-friendly fashion, and sustainable travel can help readers make more environmentally conscious decisions in their daily lives. This can be anything from using reusable shopping bags to supporting local farmers and businesses that prioritise sustainability.

Magazines also have the ability to raise public awareness of environmental injustices and disparities, particularly in disadvantaged communities [7]. Magazines that draw attention to issues like environmental racism, water pollution, and deforestation in sensitive areas can help amplify the voices of those most affected by environmental degradation. By encouraging empathy and a sense of solidarity, this can inspire readers to support marginalised communities in their fight for environmental justice and a healthy environment. Magazines can offer a platform for talking about and debating significant environmental concerns. Magazines facilitate a diversity of ideas and perspectives, thereby prompting readers to critically engage with intricate environmental issues and explore various solutions to environmental challenges. Promoting environmental practices and policies that benefit both people and the environment is easier for informed and empowered individuals to accomplish.

A few good examples include the "African Wildlife" that focuses on wildlife conservation and the protection of natural habitats in Africa, the "Environmental

Africa" that provides a platform for African environmental experts and activists to share their insights and perspectives on the environmental challenges facing the continent, and others such as the "Africa Geographic," "Conservation Africa," and "Green Africa" [7]. These magazines play a vital role in informing the public about the environmental challenges facing the continent and inspiring action to address these issues.

3.1.5 *Posters*

It is a great idea to use posters to raise awareness of environmental issues. Posters can use eye-catching imagery and concise text to educate and raise awareness of the necessity of environmental conservation [11].

Posters can be used to promote environmentally beneficial practices including recycling, using renewable energy sources, and reducing litter. People can be motivated to make more environmentally friendly decisions in their daily life by posters that highlight the benefits of sustainable living. Using posters to advocate for changes in legislation to protect the environment can prove to be a highly successful tactic. Activists can use posters urging politicians to take immediate action on environmental issues to gain public support and put pressure on legislators to pass eco-friendly laws. Posters about environmental issues can pique people's curiosity and inspire them to take action. Posters that incorporate calls to action, such as participating in a rally, submitting a petition, or volunteering for an environmental organisation, can inspire individuals to take up local activism. Moreover, posters can also be used to honour environmental achievements and anniversaries [12]. Posters that showcase accomplishments, such as the successful restoration of a natural area or the implementation of a conservation project, can inspire people to keep working towards a sustainable future.

3.1.6 *Billboards*

As a result of their broad appeal and capacity to attract attention, billboards can be placed thoughtfully in crowded areas [13]. With their striking images and impactful language, billboards can raise public awareness of significant environmental issues. The goal is to stimulate curiosity and inspire people to learn more about the issue and take action.

Moreover, organisations can promote their sustainability initiatives and show their commitment to environmental stewardship by using billboards. Companies can attract environmentally conscious consumers and inspire others to follow suit by exhibiting their efforts to reduce carbon emissions, protect water resources, or advance renewable energy (Fig. 3.3).

Billboards are a useful tool for encouraging people to live more ecologically conscious lives. Using the bus, recycling, and reducing litter are just a few of the things that billboards messaging can promote [14]. By emphasising the

Fig. 3.3 Billboard. Courtesy of Wikimedia Commons (2025)

importance of specific activities, billboards can shape attitudes and inspire positive change.

Besides, billboards can also be used to honour environmental achievements and present motivational stories about sustainability and conservation [13]. Messages can emphasise successful restoration projects, wildlife conservation initiatives, or neighbourhood projects that have enhanced the environment. By sharing these success stories, billboards have the potential to inspire individuals to take action and inspire optimism.

In fact, billboards are a powerful tool for uniting local communities around a common cause and building support for environmental issues. For example, billboards can request of people to donate to environmental organisations, sign petitions, help with clean-up projects, or go to demonstrations. Moreover, campaigners for environmental causes can use billboards' visibility and reach to successfully spread their message and bring about meaningful change.

3.1.7 *Newsletters*

Newsletters provide a venue for informing readers about several issues [15]. Organisations can help spread relevant environmental articles, infographics, and statistics to help increase understanding and awareness of the challenges facing our globe. These newsletters can also highlight achievements in environmental protection and positive tales to inspire hope and action.

Newsletters are an excellent tool for providing information and resources on living a more sustainable lifestyle [15]. Organisations can provide information on energy-saving tips, recycling laws, eco-friendly alternatives, and sustainable living possibilities. By providing readers with useful information, newsletters can encourage positive behavioural changes and the promotion of an environmentally responsible lifestyle.

Studies show that newsletters are a great tool for showcasing the outcomes of environmental and conservation efforts [16]. By sharing success stories, organisations can inspire people to become involved, recognise the work of volunteers and supporters, and celebrate triumphs. By demonstrating the achievements of environmental projects and programmes, these narratives can also help readers gain confidence and trust.

Besides, a good way to engage with the community and foster a feeling of connection and belonging is through newsletters [16]. By highlighting the experiences of neighbourhood environmental heroes, bringing attention to upcoming events, and enticing readers to participate in campaigns and activities, organisations can promote a sense of community and togetherness. This engagement can lead to increased support, involvement, and cooperation in environmental endeavours.

Newsletters can contain strong calls to action to compel readers to adopt particular environmental steps. Newsletters have the power to inspire and encourage others to become involved, whether it is by donating money, signing a petition, taking part in a demonstration, or assisting with clean-up. By providing readers with opportunities to become involved and make a difference, organisations can harness the power of their audience and have a collective impact on the environment.

3.1.8 Journal Articles

Journal articles include professional analysis, factual data, and research findings that can influence governmental decisions, raise public awareness, and promote sustainable practices, making them crucial resources for environmental communication in Africa.

The fact that journal articles go through a rigorous process of investigation and peer review, makes this type of media a trustworthy source of information on environmental issues in Africa [8]. Citing these publications in communication efforts enables people and organisations to back up their claims with reliable data, which increases audience trust. Researching how deforestation affects biodiversity in a specific area can help to emphasise the significance of conservation efforts and the requirement for sustainable land use practices.

Research shows that journal articles are great teaching tools for expanding understanding and awareness of complex environmental challenges since they are written by subject-matter experts [9]. By translating these materials into plain English, environmental communicators can help educate the public about topics like pollution, resource management, and climate change. Armed with this knowledge, people can feel more equipped to take charge, make informed choices, and support initiatives that promote environmental preservation and sustainability.

Journal articles can have a big impact on African environmental policies and legislation [8]. By presenting study results and highlighting the detrimental consequences of inactivity, environmental communicators can encourage legislative reforms that address pressing environmental challenges. A study comparing the prices of renewable energy sources and fossil fuels, for example, might be used to make a compelling case for government investment in clean energy infrastructure.

Through journal articles, researchers, practitioners, policymakers, and activists can collaborate on projects, share ideas, and build networks for environmental action in Africa. Citing relevant articles in communications allows environmental groups to build relationships with like-minded individuals and organisations, leverage information, and work together towards common goals. Collaboration of this nature can lead to innovative solutions, more clout, and a stronger voice for environmental issues across Africa.

3.1.9 Books

Books are powerful vehicles for communication that can be used in a variety of ways to inform readers, inspire action, and raise awareness of Africa's environmental problems [17].

Books can serve as educational resources to help readers understand the importance of protecting and maintaining the environment. Through thorough research and instructive content, books can provide significant insights into the various ecological problems that Africa is facing [17]. By spreading knowledge and understanding of these issues, books can help people make informed decisions and take action to preserve their environment for future generations.

Effective environmental conservation programmes and efforts in Africa can be showcased through books. Books can be a fascinating source of inspiration and encouragement for people to become involved in environmental concerns since they tell the inspiring stories of individuals, groups, and local communities. Given that positive change is achievable, these narratives can motivate readers to address issues within their own communities (Fig. 3.4).

Books can be used to communicate African traditional ecological knowledge and practices that have been handed down through the ages. By incorporating indigenous perspectives and wisdom into environmental communication, books can help close the knowledge gap between traditional knowledge and modern science and assist a comprehensive approach to environmental conservation. This can be especially useful in promoting culturally appropriate and ecologically beneficial actions that honour local traditions and customs.

Studies indicate that books can inspire activism and mobilise support for environmental issues in Africa [17]. Books can help readers develop a feeling of responsibility and solidarity by highlighting the pressing need for action and offering practical solutions. Books with compelling narratives and persuasive arguments have the power to inspire readers to become environmental justice

Fig. 3.4 Books. Courtesy of Freepik (2025)

activists. They can also exhort people to seek legislative changes that prioritise both human and environmental welfare and to speak out against injustices against the environment.

3.2 Digital Platforms

Digital platforms have become indispensable instruments for environmental communication because of their broad reach and ease of use [18]. A few of the numerous online channels that individuals and groups can utilise to effectively promote change, raise public awareness, and rally support for environmental issues are social media, websites, blogs, and online forums.

3.2.1 Social Media

Social media has become an integral part of our daily lives, with billions of people communicating with one another, sharing information, and expressing themselves on platforms like Facebook, Instagram, LinkedIn, and Twitter. Social media is often criticised for its negative effects on mental health and privacy, but it can also be a very powerful tool for communication, especially when it comes to bringing important societal issues to light, like the environment.

3.2.1.1 Facebook

Facebook has the power to inform and enlighten people [18]. Spreading information about environmental issues through articles, videos, and infographics can make people more aware of the issues facing our planet and the solutions at our disposal. Facebook can be helpful for disseminating important information and starting discussions on pressing environmental concerns since it can target specific groups and has a wide reach.

Facebook can serve as a platform for advocacy and activism. Environmental organisations and activists can create specific pages and groups to garner support for a range of causes, including as climate change, deforestation, animal conservation, and plastic pollution. By organising events, campaigns, and petitions on Facebook, activists can encourage their followers to take up causes locally or virtually.

Sharing success stories and best practices on Facebook is a good idea. People and organisations can inspire others to adopt eco-friendly practices into their daily lives by exhibiting examples of modern technology, eco-friendly living, and conservation efforts. Facebook can be a useful platform for promoting green products, showcasing eco-friendly initiatives, and offering tips on how to reduce waste so that others are motivated to follow suit.

Aside from this, Facebook also facilitates networking and collaboration amongst organisations and individuals with shared interests [19]. Environmentally conscious people can work together to accomplish common goals by sharing knowledge, resources, and first-hand encounters. By means of resource pooling, project cooperation, and forming partnerships, Facebook can facilitate the growth of a strong network of environmental advocates who can cooperate to effect change.

In addition, Facebook has the potential to inspire individuals to take action by interacting with them. Organisations that provide calls to action, volunteer opportunities, and ways to become involved can encourage individuals to take positive environmental action. Facebook can make it possible for users to actively participate in the fight against environmental degradation and climate change by allowing them to join petitions, participate in cleaning events, and support environmentally aware businesses.

3.2.1.2 Instagram

Instagram allows environmental groups and campaigners to reach a global audience with visually striking content. By posting pictures and videos of wildlife conservation efforts, environmental degradation, and sustainable living techniques, users can visually engage and inform their followers. Using visual storytelling is a great way to engage people who cannot otherwise be interested in environmental issues.

Similar-minded individuals can connect and create a sense of community around environmental problems on Instagram by using the hashtag feature. By using popular environmental hashtags such as #zero waste, #climate action, and #sustainable living, people can take part in conversations, share resources, and collaborate to solve pressing environmental issues. Additionally, by making it easier for users to access and interact with pertinent material, hashtags expand the reach and impact of environmental issues.

Instagram stories offer an even more ephemeral way to share content and communicate with followers in real time. Environmental organisations can use Instagram stories to share updates on campaigns and initiatives, provide behind-the-scenes glimpses into their work, and interact with their audience through polls and question-and-answers sessions. Viewers are urged to check back frequently for changes because stories are only available for a short while, which keeps them engaged and informed about current environmental issues.

3.2.1.3 Twitter

Twitter now called X can be used to support environmental campaigns and communications [20]. With X, environmental organisations and activists can instantly reach a global audience with vital information on conservation efforts, climate change, and other environmental issues in a matter of seconds. By using relevant hashtags and engaging with their followers, X users can inspire others to take action and spread awareness of their cause [20].

Environmental campaigners can use X to engage with interested persons and encourage conversation about environmental issues. Through X chats, live-tweeting events, and direct messaging, environmental campaigners can communicate with legislators, business executives, and local residents to discuss solutions to pressing environmental challenges. Since it allows for transparent and honest communication, X can promote consensus-building and cooperative efforts towards a more sustainable future.

X can be used to convey news and updates about environmental developments. X can be used by environmental experts and groups to share breaking news and ideas in real time, covering everything from fresh scientific research findings to legislation developments. X users who stay up to date on the latest developments can inspire action on significant environmental problems and help them make informed judgements.

Besides, using X is one approach to get support for environmental projects and campaigns. To raise awareness of their cause and garner support, environmental activists can turn to social media influencers, celebrity endorsements, and viral content. X can be used to draw attention to grassroots initiatives and inspire people to get involved in environmental activism, whether it is by organising protests or raising money.

X can be an effective tool for making companies and governments answerable for their environmental impact. By taking part in events like #FridaysForFuture and #ClimateStrike, X users can put pressure on companies to adopt sustainable practices, governments to pass stricter environmental laws, and individuals to make more environmentally responsible decisions. Through social media activism and public pressure, X can influence change at all societal levels and contribute to a more ecologically friendly and sustainable society.

3.2.1.4 LinkedIn

LinkedIn helps users connect with firms, professionals, and thought leaders in industries [18]. By building a network of like-minded individuals, people can share best practices, collaborate on initiatives that promote sustainability, and remain informed about the most recent developments in environmental concerns.

It is a great idea to provide information on environmental issues on LinkedIn in the form of articles, blog entries, videos, and other resources. By regularly sharing informative and engaging content, environmental communication can inspire people in networks to take up local and personal activism and teach them the importance of environmental conservation. In addition, LinkedIn allows people to showcase their sustainability efforts and successes. By sharing updates on green efforts, such as reducing carbon emissions, setting up recycling systems, or supporting local environmental issues, people can raise awareness and inspire others to do the same.

LinkedIn facilitates meaningful engagement with stakeholders, including investors, employees, legislators, and customers. Participating in discussions, responding to inquiries, and gathering feedback from your network can help environmental campaigners build credibility, promote collaboration and communication on environmental issues, and demonstrate environmental campaigners' commitment to sustainability. The platform can also be used to promote regional, governmental, and global policy changes in environmental communication. By sharing information, research findings, and success stories, environmental campaigners can stimulate advocacy activities, raise awareness of the need for more stringent environmental regulations, and motivate others in environmental campaigners' network to support sustainable policies.

3.2.1.5 YouTube

YouTube facilitates the dissemination of accurate and up-to-date information about societal issues [19]. Experts can use video content to share their knowledge and research on environmental issues with a wider audience since it breaks complex topics down into easily comprehended facts. YouTube can be used to share environmental success stories and inspire people to take action. Stories of individuals and groups taking steps towards sustainability and conservation can motivate others to take action in their own lives, according to content creators. This can entail highlighting original approaches to environmental problems, supporting eco-friendly products and practices, and recognising those who are going above and beyond in the fight against environmental degradation.

YouTube can also be used to raise awareness of environmental issues and motivate people to take action. By incorporating calls to action into their videos, video producers can encourage their viewers to support environmental causes, participate in conservation initiatives, and advocate for legislative changes. Environmental campaigners could consequently come into contact with a group of people who are as passionate about protecting the environment and promoting sustainable development as they are.

YouTube can be a valuable resource for organising grassroots movements and promoting environmental action. When content creators provide information about upcoming events, marches, and campaigns, they can connect with viewers who are interested in becoming involved in environmental activism. This might give campaigners a stronger voice and give people the courage to stand out for the environment in their communities.

In addition, environmental campaigners can use YouTube as a medium for educating people about the environment. Content creators might potentially raise public awareness of the need of preserving the environment for future generations by creating engaging and informative content on topics like conservation, biodiversity, and renewable energy. People can be inspired to live more sustainably and to take better care of the environment as a result.

3.2.1.6 Blogs

One of the most crucial functions that blogs perform in environmental communication is educating and informing the public about key environmental problems [19]. By disseminating news articles, research findings, and instructional materials regarding significant environmental issues, bloggers can aid in the process of raising public awareness. By presenting this knowledge in an easy and engaging way, blogs can help educate a big audience and inspire action.

Blogs are an essential tool for environmental communication since they promote sustainable lifestyle choices and behaviours. Bloggers can use their platforms to spread advice and suggestions on how individuals can reduce their impact on the environment by, for instance, consuming less energy, producing less garbage, and patronising green businesses. Blogs can inspire readers to make positive changes in their own lives by highlighting the benefits of sustainable living and providing practical guidance on how to put these changes into practice.

Besides, blogs can be used to advocate for changes to regional, national, and international policies in addition to educating the public and promoting sustainable practices. By highlighting the need for action, addressing significant environmental problems, and researching relevant laws, bloggers can help mobilise support for tighter environmental regulations and actions. A blog's ability to engage the public, activists, and policymakers can help to amplify the voices of those who are advocating for positive change. Blogs can be used to encourage environmental action and a sense of community. Bloggers can create a platform for discussion, networking, and collaboration amongst groups and people who have similar goals and interests. By sharing success stories, organising events, and encouraging group action, blogs can inspire individuals to become involved in environmental activism and make improvements in their local communities.

3.2.1.7 Websites

Websites that provide up-to-date information on facts, figures, and research findings can educate the public about the state of the environment and the possible repercussions of doing nothing [18]. This information can raise awareness and motivate individuals to take action for environmental conservation.

Websites can be used to showcase success stories and best practices for environmental protection. Websites that highlight the initiatives of individuals, communities, and organisations that are transforming the world can inspire others to do the same. Positive reinforcement has the potential to accelerate environmental conservation initiatives by motivating more people to become involved. Websites can encourage discussion and debate around environmental issues.

Moreover, websites can enable users to share their thoughts, ideas, and experiences concerning the environment might provide a forum, comment sections, social network integration, and other features. By feeling more linked to one another and collaborating, people who are passionate about environmental conservation can gain from this information and opinion exchange. Websites can be used to promote environmental problem advocacy and action.

In addition, websites that provide tools like contribution links, event calendars, petitions, and volunteer opportunities can inspire users to take proactive steps to save the environment. By turning knowledge into action, this call to action can spur real environmental change.

3.2.1.8 Podcasts

Podcasts provide a platform for experts to share their knowledge and viewpoints on environmental issues. Podcasts that feature interviews with scientists, academics, activists, and government representatives can educate listeners about significant environmental challenges and potential solutions. Listeners can get a greater understanding of complex environmental concerns through these informative presentations, empowering them to make informed decisions daily.

Podcasts can be utilised as a storytelling tool to show how environmental issues impact individuals personally. By presenting the real-life stories of those affected by environmental problems, podcasts can foster empathy and inspire action. The emotional resonance these stories have with their audience motivates individuals to support environmental concerns and make positive behavioural changes.

Moreover, podcasts have the power to inspire community engagement and action about environmental issues [19]. With lively discussions, live events, and audience participation, podcasts can help individuals who are passionate about environmental conservation feel more connected to one another. Listeners can unite to plan clean-up events, back sustainable projects, and advocate for legislative changes to forge a powerful collective voice for environmental action.

In addition, podcasts offer a creative and engaging medium for showcasing innovative solutions to environmental issues. Podcasts can inspire listeners to embrace eco-friendlier activities by disseminating tales of green businesses, green technologies, conservation campaigns, and sustainable methods. Through turning motivation into action, listeners can discover practical methods for reducing their carbon footprint, protecting the environment, and preserving wildlife.

3.2.1.9 Online Forums

Online forums provide a platform for the sharing of knowledge and expertise on environmental issues between activists, experts, and concerned citizens [19]. This can help dispel misconceptions and misunderstandings about environmental issues, educate people about sustainable practices, and raise awareness of how serious environmental problems are. Engaging in online discussion boards and material exchanges can help people become more knowledgeable about environmental issues and inspire others to take action.

Besides, online forums can be used to unite supporters of environmental causes and promote a sense of solidarity among like-minded individuals. Finding others who share their concerns and values can help individuals organise events, work on campaigns, and advocate for legislative changes that protect the environment. This sense of unity and purpose can inspire people to take action and better their communities.

Online forums can also help persons and organisations working on environmental issues network and collaborate. By building relationships with authorities, potential partners, and other stakeholders, those working in the environmental sector can exchange ideas, pool resources, and collaborate on projects that have a greater impact than they could alone. In the context of the environment, networking and collaboration can foster creativity, create alliances, and provide possibilities for learning and growth on both sides.

Moreover, online forums can serve as useful spaces for environmental action and advocacy. People and organisations can share petitions, arrange protests, raise money, and mobilise support for programmes that aim to protect the environment and hold polluters accountable using online forums. By using the collective power of online forums, people can raise awareness of environmental injustices; fight for meaningful change at the local, national, and worldwide levels; and magnify their voices.

3.2.1.10 Virtual Events

Virtual events can reach a global audience in a way that is not feasible for live events [20]. Environmental organisations can reach a global audience by hosting webinars, seminars, and virtual conferences. This promotes change on a larger scale and makes the sharing of information easier. As a result of this accessibility, people who might not be able to attend in-person events can still engage and have important conversations about the environment.

Virtual events include a variety of interactive components that might increase participant involvement. These interactive elements, which include virtual tours of natural settings and in-person question-and-answer sessions with experts, can deliver users a more personalised and engaging experience. Environmental organisations can use virtual reality and gamification to boost understanding about environmental issues. As a result, learners can be more engaged and will understand the subject matter more deeply.

Besides, planning virtual events can be a cost-effective and ecological alternative to traditional live events. If organisations can eliminate the need for travel, hotel, and physical venue rents, they can save costs and their carbon impact. This enables greater funding to be allocated to deserving environmental campaigns and initiatives, which ultimately can lead to an improved outreach and communication plan over the long run.

In addition, virtual events provide insightful information that can help businesses evaluate the success of their initiatives and improve their communication strategies. Metrics like attendance rates, engagement levels, and audience comments can help environmental groups identify areas for improvement and get critical insights into the effectiveness of their goal. This data-driven approach can help ensure that communication efforts are efficient and targeted by using data to shape next campaigns and events.

3.2.1.11 Mobile Applications

With smartphones and other mobile devices becoming more and more popular, mobile applications have the potential to educate a broad audience about the need of environmental preservation [19]. Interactive features such as movies, infographics, and tests can teach users more about how their activities impact the environment.

Mobile applications make it simple to get information about sustainable practices and eco-friendly products [21]. These applications can help people make positive changes in their daily lives by providing them with the knowledge and tools necessary to make environmentally responsible decisions. With knowledge on green energy options and recycling regulations, mobile applications can help users reduce their carbon footprint and embrace sustainable practices.

With the use of mobile applications, users can monitor their impact on the environment and their progress towards sustainability objectives. The features of these programmes, such as carbon footprint calculators, waste reduction guidance, and energy consumption monitors, can help users make informed decisions and track their progress over time. When users receive this real-time feedback, they can be motivated to take action and make major behavioural changes.

Mobile applications can help with community involvement and collaboration on environmental projects. These applications have the potential to promote a sense of community around environmental issues by connecting users with like-minded individuals and groups. With the use of resources such as forums, event calendars, and volunteer opportunities, users can work together to save the environment and encourage sustainable practices.

In addition, mobile applications can be used as advocacy tools by environmental organisations and campaigners. By providing a platform for users to exchange news, updates, and calls to action, these applications have the potential to mobilise supporters and raise awareness about pressing environmental issues. Social network integration, donation portals, and petitions are examples of technologies that enable individuals to immediately get engaged in advocacy activity and support organisations that align with their values.

3.2.1.12 Artificial Intelligence

Artificial intelligence can understand people's preferences, actions, and interests by processing massive amounts of data [5]. Using this information, environmental organisations can create customised messaging that speak to specific individuals and inspire them to take action to protect the environment. Whether recycling, reducing energy consumption, or promoting sustainable practices are the objectives, artificial intelligence can tailor messages to improve engagement and motivate behaviour change.

To provide intelligent analysis of environmental issues, artificial intelligence can gather and evaluate data from a variety of sources, such as sensors, social media, and satellite imagery. Artificial intelligence has the ability to convert complex data into visually intuitive representations that effectively communicate the urgency and gravity of environmental challenges [19]. Interactive maps, graphs, and infographics presenting data on air pollution, deforestation rates, and climate change estimations can be produced via artificial intelligence -driven systems.

Chatbots powered by artificial intelligence can serve as virtual assistants, providing information, assistance, and advice on environmental concerns in real time. Users can ask chatbots immediate questions on sustainable methods, animal welfare, and environmental laws. By offering easily accessible and engaging environmental information, chatbots can empower individuals to make informed decisions and encourage environmental conservation.

In addition, artificial intelligence systems can predict future environmental trends and outcomes based on historical data and current patterns [14, 16]. Using predictive modelling, environmental organisations can forecast the effects of pollution, biodiversity loss, and climate change on ecosystems and populations. Plans, strategies, and campaigns to address emerging environmental concerns before they spiral out of control can be developed utilising these results.

3.2.1.13 Virtual Reality

With virtual reality, users can experience first-hand the effects of environmental degradation since it provides immersive experiences that replicate real-life scenarios. For example, users can visually explore a deforested area to see how biodiversity and ecosystems are impacted by logging. By immersing users in these experiences, virtual reality has the potential to evoke powerful emotions and highlight the need of environmental conservation.

Scholars posit that one powerful educational tool that can be used to teach people about environmental issues in an engaging and interactive fashion is virtual reality [18]. For example, virtual reality simulations can show how different regions of the world are being affected by climate change or promote eco-friendly practices like using renewable energy. Engaging with these simulations through virtual reality can help people get a greater understanding and knowledge of environmental challenges.

By creating virtual tours of natural ecosystems, national parks, and conservation areas utilising virtual reality, users can engage with these environments in a realistic and immersive way. For example, users can digitally visit an affected place and observe the threats this fragile ecosystem confronts. Additionally, virtual reality campaigns can be used to raise awareness about specific environmental issues like plastic pollution and deforestation by creating interesting and instructive interactive experiences for consumers.

Besides, virtual reality holds great potential as a tool for environmental activism and fundraising. Organisations that create virtual experiences that highlight the importance of environmental conservation might inspire individuals to take action and support their cause. Virtual reality experiences have the potential to showcase success stories, accentuate the efforts of conservation groups, and illustrate the impact of donations on environmental initiatives [18]. By using virtual reality to tell compelling stories and establish an emotional connection with customers, businesses can mobilise support for their environmental projects.

3.3 Successful Campaigns for Environmental Communication

Environmental communication campaigns are crucial for increasing public awareness and spurring action for environmental preservation. Across Africa, there have been several successful campaigns that have spurred conservation projects and positive change.

3.3.1 Trashy Bags

An innovative Ghanaian company called Trashy Bags has been tackling the issue of plastic waste head-on with its environmental awareness campaign [22]. Trashy Bags was established in 2008 with the goal of enhancing Ghana's environmental circumstances via the collection and transformation of plastic waste into fashionable bags and accessories.

The environmental education campaign by Trashy Bags is a useful tool for spreading awareness of the negative effects that plastic trash has on the environment and encouraging sustainable lifestyles in both individuals and communities. To mitigate plastic trash's detrimental impacts on the environment, the campaign aims to increase public awareness of the importance of recycling, reusing, and reducing plastic waste.

Trashy Bags is a vital component of environmental sustainability since it creates jobs for those living in impoverished regions while simultaneously collecting and recycling plastic waste [23]. Trashy Bags uses recycled materials to create fashionable bags and accessories that not only help local communities but also fight plastic pollution.

A variety of environmental communication initiatives are being carried out by Trashy Bags, including social media campaigns, neighbourhood outreach initiatives, and educational seminars. These initiatives aim to mobilise public support for environmental protection and the decrease of plastic waste [22].

One of the shortcomings of the Trashy Bags programme in Ghana is its restricted impact and reach. Their views about how crucial it is to recycle and lessen plastic waste are not being heard by many people. This can be the consequence of a lack of financing for marketing and public relations, as well as a failure to collaborate with other environmental organisations and governmental bodies. Without extending its reach, Trashy Bags is finding it impossible to have a significant impact on Ghana's plastic pollution issue.

3.3.1.1 Lessons

Trashy Bags' inventive way of upcycling plastic waste into stylish and functional bags has caught the attention of shoppers both locally and internationally [24]. Since Trashy Bags uses waste materials into stylish products, this company has attracted a loyal following of customers who value sustainability and ethical production.

Trashy Bags has a big social impact since it provides work opportunities for poor people in Ghana. Many of the young, low-income men and women who work at their production sites are local craftsmen. In addition to helping the environment, Trashy Bags creates cash for those who most need it by paying fair wages and fostering skill development.

Moreover, Trashy Bags has a strong brand and a good reputation in the sustainability sector. They have received attention and awards from international media outlets for their commitment to environmental preservation and social effect. With this recognition, Trashy Bags has become a leader in Ghana's fight against waste pollution and has contributed to raising awareness of the issue of plastic pollution.

3.3.2 The Green Belt Movement

The Green Belt Movement founded by Wangari Maathai, a Kenyan environmentalist and political activist, urges communities—women in particular—to plant trees as a means of preserving the environment and supports sustainable development. The movement has played a major role in Kenyan and global environmental conservation, and it still has an impact on other countries [25].

One of the objectives of the Green Belt Movement is to increase public awareness of the need for environmental conservation. The movement raises awareness of the detrimental effects of soil erosion, deforestation, and climate change in communities by planning planting activities, workshops, and lectures. These efforts motivate people to take action to protect the environment for future generations.

Rural women's empowerment is also greatly aided by the Green Belt Movement initiative [21]. Women have a strong stake in environmental preservation since they are often the primary caregivers and providers in these communities. The organisation provides training in sustainable agriculture, tree-planting, and leadership to assist women in becoming environmental stewards in their local communities. By including women in environmental preservation, the movement promotes gender equality and socioeconomic development.

Furthermore, the Green Belt Movement campaign advocates for governmental modifications that support environmental preservation efforts [8]. The movement has been successful in pushing for legislation controlling logging, land use, and water management to protect natural resources. The campaign ensures that environmental issues will remain at the top of the national agenda by actively engaging with legislators and stakeholders.

One of the problems with the Green Belt Movement campaign is its lack of funding and resources [8]. The Green Belt Movement is a grassroots organisation that primarily receives funding for its conservation efforts from donations and grants. This often leads to constraints in terms of implementing larger-scale initiatives and reaching more marginalised communities. Insufficient funding might hinder the campaign's ability to expand and have an impact, which would lower its overall effectiveness in promoting environmental preservation

Another shortcoming of the Green Belt Movement is its concentrated leadership structure. Even though Wangari Maathai the founder was an inspiring and strong leader, her passing in 2011 left a void in the group. Internal strife and problems resulting from the lack of a strong and unified leadership after her death have affected the campaign's overall efficacy [8].

3.3.2.1 Lessons

One of the Green Belt Movement's best features is its grassroots approach to environmental conservation. Through direct community empowerment and participation, the Green Belt Movement has built a strong feeling of ownership and devotion to conservation efforts. Long-lasting and robust are the environmental benefits of this bottom-up approach.

The Green Belt Movement's primary strength is its emphasis on education and capacity-building [21]. In addition to planting trees, the initiative provides training in sustainable land management practices and teaches communities the importance of environmental preservation. By enhancing their capacity, the Green Belt Movement ensures that local communities have the skills and knowledge necessary to continue conservation activities long after the campaign is over.

The global public's awareness of significant environmental issues has also been effectively raised by the Green Belt Movement. The Green Belt Movement has effectively influenced policies and promoted sustainable development practices across numerous industries through its lobbying efforts and partnerships with other institutions. By drawing attention to important environmental issues, the Green Belt Movement has played a crucial role in shaping the global conversation on sustainability and conservation.

3.3.3 The Plastic Agenda Campaign

The Plastic Agenda Campaign was launched in South Africa as a reaction to the growing environmental crisis caused by plastic waste [26]. Through encouraging sustainable practices and raising public awareness of the damaging effects of plastic waste on the environment, this initiative seeks to reduce plastic waste and consumption.

The primary objective of the Plastic Agenda Campaign is to increase public awareness of the harm that plastic trash does to animals, humans, and marine life. Plastic waste can be found in landfills, rivers, and oceans. This pollution entangles and ingests marine life, contaminates food and water sources, and releases dangerous chemicals into the sky. By highlighting these detrimental impacts, the campaign aims to convince individuals and institutions to move away from single-use plastics and towards more ecologically friendly alternatives.

Another role of the Plastic Agenda Campaign is its lobbying for national legislation and policies that address plastic pollution. South Africa has already started tackling this issue by imposing a fee on plastic bags and banning some types of plastic products. There is still more work to be done, though, to effectively manage plastic waste and develop a circular economy that has a less negative impact on the environment. The campaign works with businesses, civil society organisations, and government officials to promote stronger laws and initiatives that prioritise sustainability and resource efficiency.

The Plastic Agenda Campaign not only advocates for legislative reforms and raises awareness of the issue but also promotes community engagement and action to decrease plastic pollution [27]. The initiative promotes responsible plastic usage by organising recycling drives, educational seminars, and clean-up events, encouraging everyone to play a proactive role in reducing plastic waste. The campaign also promotes working together with local businesses, educational institutions, and communities to create a network of individuals committed to advancing environmental preservation and sustainability.

Nevertheless, only a tiny percentage of South Africa's population—mostly educated urbanites—are being reached by the Plastic Agenda Campaign. Plastic pollution affects all people, regardless of their financial status or level of education, making it a serious issue. By failing to engage a larger audience, the campaign bears the risk of perpetuating the misconception that plastic waste is not a significant issue for all South Africans.

The Plastic Agenda Campaign is struggling to make compelling calls to action that would motivate individuals and groups to take decisive action to cut down on plastic waste. Raising awareness is important but so is providing tools and practical solutions to enable people to make changes in their daily lives. If the campaign does not include clear directions on how people can reduce their use of plastic or participate in cleaning efforts, individuals can feel overwhelmed and powerless to effect change.

3.3.3.1 Lessons

One of the Plastic Agenda Campaign's greatest assets in South Africa is its ability to unite and include a wide range of stakeholders, including governmental organisations, charities, businesses, and private individuals [28]. By combining various groups around the same goal of fighting plastic pollution, the project promotes cooperation and cooperative action, which leads to better solutions and more lasting impacts.

Another plus for the Plastic Agenda Campaign is its concentration on raising public awareness and educating people about the negative environmental impacts of plastics. The campaign uses targeted messaging, outreach programmes, and educational materials to provide individuals the information they need to make informed decisions about how much they consume and how to manage their trash. By providing resources and information that raise public awareness of the issue of plastic pollution, the initiative encourages people to modify their behaviour and take action to reduce their plastic footprint.

Additionally, the Plastic Agenda Campaign uses narrative and advocacy to inspire and drive change. Through the sharing of personal accounts, testimonies, and case studies, the campaign highlights the impacts of plastic pollution in the real world and encourages its audience to take action and support sustainable solutions. This strategy increases the audience's empathy and sense of urgency. Using story, the campaign humanises the issue of plastic pollution and builds an emotional bond with the public, raising the possibility that people would be motivated to take up new habits and support legislation that prioritises the sustainability of the environment.

Utilising several venues and communication methods, the Plastic Agenda Campaign in South Africa engages a broad audience and disseminates its message. The effort meets people where they are and initiates crucial conversations on plastic pollution using a range of strategies, such as social media advertising, public events, community outreach programmes, and educational seminars. By mixing traditional and digital media, the campaign is reaching a wider audience and sparks a movement for change that transcends geographic and demographic borders.

3.3.4 Clean Up Kenya

Clean Up Kenya started out as a grassroots initiative in 2002 with the intention of raising public awareness of the harmful consequences of littering and inspiring people to take action to improve the environment and make it cleaner and healthier [29]. Through a series of organised clean-up events around the country, the campaign has mobilised thousands of volunteers to help with the clean-up of public spaces and raise awareness of the need of proper waste management [30].

One of the campaign's shortcomings is the overall lack of public awareness and participation in Clean Up Kenya. Many organisations and groups support the campaign, yet a significant part of Kenyans are still unaware of it or the need for environmental conservation. This ignorance can limit the campaign's effectiveness and have a negative influence on the environment.

Another problem with the Clean Up Kenya effort is that it lacks finances and long-term funding. The campaign mostly relies on contributions and volunteers to carry out its operations, which can be challenging to sustain over time. Without enough funding and resources, the campaign is struggling to reach a larger audience and implement more ambitious plans to address Kenya's environmental problems.

3.3.4.1 Lessons

Despite these shortcomings, several qualities of the Clean Up Kenya initiative have contributed to its success. One of the campaign's advantages is the amazing support and engagement from the community. The effort to participate in environmental conservation and clean-up programmes has motivated communities around Kenya, strengthening links between neighbours and fostering collaboration.

An additional advantage of the Clean Up Kenya initiative is its emphasis on education and awareness-raising. The campaign has developed educational projects and programmes to promote environmentally conscious behaviour and raise public awareness of environmental issues. Community members can feel more accountable and invested in the campaign if they are made aware of the importance of environmental conservation.

In addition, Internet and social media platforms have been effectively integrated by the Clean Up Kenya campaign to engage supporters and expand its reach. Through social media campaigns and online activities, the movement has been able to raise awareness, recruit volunteers, and lobby for policy reforms to address environmental problems in Kenya.

The Clean Up Kenya campaign has established solid partnerships with governmental bodies, non-profits, and corporate partners in order to collaborate on environmental conservation projects and initiatives. By working together with a wide range of stakeholders, the campaign can make better use of networks, resources, and expertise to meet its goals and have a greater overall environmental impact.

3.3.5 The Green Legacy Initiative

The Green Legacy Initiative (GLI) commenced in June 2019 with the aim of creating a sustainable and climate-resilient Ethiopia. The aim was to produce 20 billion seedlings over a four-year span. The Ethiopian government states that by engaging over 20 million residents nationwide, Ethiopia has successfully planted 25 billion seedlings at the end of the fourth year [31].

The Initiative aims to fulfil several goals, which might be considered a notably innovative attribute overall. This initiative favourably influences several interconnected goals, such as environmental preservation, the restoration of exhausted and impaired natural resources such surface soil and water, and the mitigation of desertification.

Ethiopia's ongoing dedication to a holistic strategy for tackling climate change and environmental degradation is shown via programmes that include agroforestry, forest sector development, urban greening and regeneration, and the integrated management of water and soil resources. The initiative is expected to have aided Ethiopia in meeting its obligations under international agreements, including the Paris Climate Change Agreement, the 2030 Agenda for Sustainable Development, and Agenda 2063. Despite criticism, the Ethiopian government asserts that The Green Legacy Initiative, aimed at alleviating the effects of climate change and deforestation in Ethiopia, has included over 25 million persons in the planting of 25 billion tree seedlings [32].

The Ethiopian Government's Green Legacy Initiative, which corresponds with several 2030 Agenda priorities, aims to improve the nation's food security. In 2022, the government declared the cultivation of over 500 million seedlings, including papayas, mangoes, avocados, and apples, which have substantial value in both domestic and global markets. This promptly supports the current effort to achieve food self-sufficiency by fostering sustainable agriculture, as delineated in Sustainable Development Goal 2. This effort aims to meet Goal 13 of the SDGs by aiding Ethiopia in minimising its vulnerability to recurring floods, droughts, and locust infestations, which illustrate the severe climatic events impacting the country. The average annual temperature in Ethiopia is projected to have increased by 0.37 degrees Celsius during the last four decades.

Despite its progress in environmental preservation and sustainability, Ethiopia continues to face challenges while implementing green economy initiatives. One of the most pressing challenges is a lack of funding and investment in green projects [32]. Many of the projects have a high initial cost, which may be a barrier to implementation. Furthermore, capacity development and technical understanding are necessary to effectively implement green economy policies and programmes.

3.3.5.1 Lessons

The Green Legacy Initiative shows that cooperation and collaboration are critical to attaining environmental preservation and sustainability. Ethiopia has successfully implemented green economy initiatives in conjunction with a diverse range of partners, including government agencies, non-governmental organisations, and private-sector businesses. This collaborative approach has helped to mobilise resources and talents to effectively address environmental challenges.

In addition, the initiative shows that community involvement and engagement are crucial to promoting environmental preservation and sustainable practices. Local communities in Ethiopia have actively engaged in afforestation, soil conservation, and water management initiatives. Ethiopia has achieved long-term success by encouraging people to take ownership of environmental efforts.

Finally, the Green Legacy Initiative demonstrates the importance of education and awareness-raising in promoting environmental preservation and sustainable practices. Ethiopia has implemented environmental education activities and campaigns to raise awareness about the need of protecting natural resources and battling climate change. Ethiopia has promoted a culture of sustainability and conservation by teaching the general public about environmental issues.

3.3.6 *The Eco Warriors of Ghana*

The Eco Warriors Ghana is a youth-led environmental organisation established in 2014 with the mission of advocating for sustainable environmental practices and combating climate change in Ghana [33, 34]. Through their social media campaigns, community outreach initiatives, and educational projects, Eco Warriors Ghana has effectively motivated young people throughout Ghana to take action on environmental problems. The organisation's actions have increased awareness of the value of environmental conservation and sparked positive reforms in neighbouring areas [35].

The Eco Warriors Ghana campaign's ambiguous aims and objectives are one of its drawbacks. The organisation's main goal is to conserve the environment, but setting specific, measurable goals is essential for tracking progress and evaluating the campaign's success. Without clearly defined objectives, it is difficult to assess whether a given organisation is making a meaningful contribution to the fight against environmental degradation.

Another shortcoming of the Eco Warriors Ghana programme is the lack of funding and resources. Like many other non-profits, Eco Warriors Ghana is funded in part by donations and grants. The group is struggling to organise events, carry out sustainability projects, or engage with its target audience without enough funds. The group needs to be able to get extra funding and resources to achieve its goal.

3.3.6.1 *Lessons*

Despite these shortcomings, Eco Warriors Ghana is nonetheless a strong force for Ghanaian environmental conservation because of several crucial characteristics. One of the organisation's strengths is its ability to inspire and involve local communities [36]. Eco Warriors Ghana's direct engagement with local leaders and communities can inspire individuals to take action and raise public awareness of environmental issues. This grassroots approach is necessary to create long-lasting change and build a more sustainable future.

An additional strength of the Eco Warriors Ghana initiative is its focus on education and capacity-building. Through training sessions, workshops, and educational programmes, the group educates the community members about conservation strategies, sustainable practices, and the importance of environmental protection. By equipping people with the skills and information needed to make positive changes, Eco Warriors Ghana is fostering the growth of a more environmentally conscious community.

Additionally, Eco Warriors Ghana has a strong online presence and a substantial social media following, which aid in the group's ability to reach a wider audience and spread its message beyond borders. By using social media platforms to share news, case studies, and educational resources, Eco Warriors Ghana hopes to inspire and motivate people all over the world to take up the fight against environmental degradation.

3.3.7 *Save the Seas of Seychelles*

To raise awareness of the critical need to preserve the fragile marine ecosystems of the Seychelles islands, the marine conservation initiative Save Our Seas Seychelles was launched in 2016 [37]. To promote sustainable practices, the programme collaborates with nearby businesses, governmental institutions, and communities. Its subjects include the devastation of coral reefs, overfishing, and plastic pollution. Save Our Seas Seychelles has successfully engaged stakeholders in public awareness campaigns, beach clean-ups, and marine education programmes to save the country's marine biodiversity [38].

One of the shortcomings of the Save Our Seas Seychelles campaign is its lack of targeted messaging. The advertisement effectively conveys the importance of ocean conservation to the public, but it might benefit from more focused content that targets key audiences. One possible focus of the campaign can be to educate the people of the Seychelles about the damage that plastic garbage does to their beaches and marine life. If the campaign's rhetoric is more precisely targeted at certain demographics, it can motivate major change and have a greater impact.

Another problem with the Save Our Seas Seychelles initiative is its limited outreach. Despite the campaign's success in reaching a substantial audience via traditional and social media means, many people are still ignorant of it or its goals. If the campaign is to make a big difference, it must reach more people and engage with a wider range of stakeholders, such local businesses, educational institutions, and community organisations. By working with these groups, the campaign can motivate more people to take action and have a big influence on ocean conservation.

3.3.7.1 Lessons

Despite its shortcomings, the Save Our Seas Seychelles campaign is a powerful force for change due to a number of crucial characteristics. One of the campaign's best features is the dedicated group of volunteers and supporters who are dedicated to making a difference in the world. These volunteers are committed to educating the public, organising cleaning efforts, and supporting legislative changes that will protect our oceans for future generations. Their seal and dedication have played a major role in the campaign's success.

Another advantage of the Save Our Seas Seychelles campaign is its strong partnerships with local, national, and international organisations. Working with organisations like the United Nations Environment Programme, the Seychelles Ministry of Environment, and local non-governmental groups can help the campaign make a bigger difference. The campaign can reach a wider audience, obtain funding and resources, and collaborate with partners on significant initiatives that will rescue our seas thanks to these relationships.

3.3.8 Greening Egypt

Egypt launched Greening Egypt, a national environmental communication programme, in 2018 to promote environmentally conscious development and sustainable growth [39]. The programme addresses issues including waste management, air pollution, and water shortages by working with businesses, civil society organisations, and government agencies to raise awareness and encourage action. Through its public awareness campaigns, educational initiatives, and lobbying efforts, Greening Egypt has made tremendous progress in promoting sustainable practices and protecting the country's natural resources [2].

One of the shortcomings of the Greening Egypt effort is the lack of resources. Adopting sustainable methods and planting trees requires a significant financial investment. As a result of insufficient financing, the programme is struggling to meet its goals. Sustaining the initiative and ensuring its success requires sufficient funding from the government and other interested parties.

3.3.8.1 Lessons

Despite shortcomings, the Greening Egypt initiative has some positive aspects that make it a noteworthy and important programme. One of the initiative's advantages is that it places a heavy emphasis on stopping deforestation and desertification. By planting trees and promoting sustainable activities, the programme aims to improve air quality, protect and preserve the environment, and develop a more sustainable ecosystem. These programmes might improve the overall health of the country and have a long-term impact on the environment.

The Greening Egypt effort is its capacity to stimulate employment and the economy. Planting trees and implementing sustainable practices can lead to jobs in the forestry and environmental sectors that support many people's lives and promote economic growth. By providing funds for green initiatives, the government can promote sustainable growth and create a more resilient and equitable economy.

The Greening Egypt initiative is also inspiring people to take action to protect the environment by raising public awareness of environmental issues. By means of organising public outreach programmes, educational campaigns, and tree-planting activities, the programme aims to engage the public and provide them with the necessary skills to make beneficial environmental impacts.

3.4 Indigenous Individual Environmental Advocates

Many Africans play an essential role as environmental stewards, protecting their land, wildlife, and natural resources. These people have a close connection to the land and understand the need of preserving it for future generations. By advocating for sustainable practices and conservation measures, they have helped to maintain ecological balance and ensure the survival of endangered species.

3.4.1 Ken Saro-Wiwa

Ken Saro-Wiwa was a Nigerian writer and environmental activist [24]. He fought relentlessly to protect the environment and promote sustainable practices. He was a strong advocate for the rights of Nigeria's Ogoni people, who were disproportionately impacted by oil extraction activities in their region.

Saro-Wiwa believed that it was important to protect the environment for future generations, and that sustainable practices were essential to ensure the well-being of both people and the land.

In response to the environmental harm caused by oil drilling, Saro-Wiwa founded the Movement for the Survival of the Ogoni People (MOSOP) in 1990 [31]. Saro-Wiwa used MOSOP to advocate for the Ogoni people's rights and seek restitution for the damage caused by oil exploration and drilling in their communities. He underlined the need of holding corporations accountable for their actions and advocating for environmental justice.

Saro-Wiwa believed that multinational firms had a responsibility to respect local communities' rights while limiting their environmental impact. Saro-Wiwa battled tirelessly to raise awareness about the environmental devastation caused by oil extraction in the Niger Delta and to get justice for the Ogoni people (Fig. 3.5).

Saro-Wiwa, on the other hand, had a number of challenges in his efforts to protect the environment and promote sustainable practices. He was opposed by powerful forces, including the Nigerian government and oil firms, who sought to silence his voice and undercut his work. Saro-Wiwa faced death threats and was eventually imprisoned and executed in 1995 on false charges [32]. Despite these obstacles, Saro-Wiwa's memory continues to motivate others who fight for environmental justice.

3.4.1.1 Lessons

Saro-Wiwa's life and activism highlights the strength of grassroots activism. He illustrated how individuals and communities can make a difference via grassroots action and advocacy. By mobilising the Ogoni people and raising awareness about environmental issues in the Niger Delta, he was able to affect change and hold the oil companies and the government accountable.

His life highlights the importance of environmental protection. Saro-Wiwa's actions highlighted the critical need to preserve the environment and natural

Fig. 3.5 Image of Ken Saro-Wiwa. Courtesy of Noo Saro-Wiwa (2025)

resources for future generations. His endorsement of sustainable approaches and smart resource management serves as a reminder to prioritise environmental conservation and sustainability.

Besides, Saro-Wiwa demonstrates the value of advocacy and civil society. He emphasised the essential role that advocacy groups and civil society organisations play in furthering social justice and holding governments and corporations accountable. MOSOP enabled him to elevate the voices of the Ogoni people and struggle for their rights on a global scale.

He also shows the courage to speak truth to power. Saro-Wiwa's brave advocacy and willingness to confront powerful interests despite significant personal risk illustrates the strength and determination necessary to expose injustice and fight for change. His legacy continues to inspire activists and environmentalists throughout the world to fight for their beliefs and strive for a fairer and sustainable future.

3.4.2 *Wangari Maathai*

Wangari Maathai was a well-known Kenyan environmentalist and political activist who relentlessly advocated for environmental preservation and sustainable values. She started the Green Belt Movement, an organisation that promotes tree planting, conservation, and women's rights [40–42].

Throughout her career, she faced several challenges, but she remained dedicated to her mission of environmental preservation and human empowerment. Her main message was on the need of safeguarding the environment for future generations. She believed that it was vital to act now to ensure that our planet could sustain life in the long term. Through her involvement in the Green Belt Movement, she motivated people to plant trees, safeguard natural resources, and advocate for sustainable development (Fig. 3.6).

Maathai also emphasised the link between environmental degradation and social issues including poverty, food insecurity, and gender inequality. She recognised that many people throughout the world rely on natural resources for a livelihood, and that environmental degradation might have serious ramifications for their health. She hoped that by addressing environmental issues, she might assist people in these communities to live better lives.

Despite her zeal and commitment to environmental action, Maathai faced various challenges during her career. She often encountered hostility from government officials, businesses, and other institutions who opposed her initiatives. She faced threats to her safety and was arrested and imprisoned for her activism. She stayed persistent in her commitment to her cause, advocating for environmental preservation and sustainability.

3.4.2.1 Lessons

Maathai's actions offer us countless valuable lessons. She illustrated how grassroots movements can impact change. She was able to make a significant impact on environmental issues in Kenya and throughout the globe by organising people in local villages.

She exemplified the importance of perseverance and resilience in the face of suffering. Despite several hurdles, she remained dedicated to her cause and never gave up on her goals.

Fig. 3.6 Image of Wangari Maathai. Courtesy of Wanjira Mathai (2025)

Besides, Maathai's taught the world the importance of education and empowering women in environmental activism. She believed that educating communities about environmental issues and encouraging women to take on leadership roles were critical to achieving long-term progress.

Maathai also illustrated the interconnection of environmental, social, and economic concerns, underlining the need of a multifaceted approach to sustainable development.

3.4.3 *Vanessa Nakate*

Vanessa Nakate, a young environmental activist from Uganda, promotes sustainable practices and environmental protection [43]. Nakate's activism has helped to raise awareness in Uganda about the need of environmental preservation and sustainability, since the country faces a number of environmental challenges like as deforestation, pollution, and climate change.

Her message on environmental protection and sustainable practices is based on the idea that everyone has a role to play in preserving the environment for future generations [43]. She believes that individuals, communities, and governments all have responsibility for minimising environmental degradation in Uganda and across the world. Nakate often highlights the need of reducing waste, adopting renewable energy sources, and protecting forests and wildlife (Fig. 3.7).

Despite her love for environmental awareness, Nakate has faced several challenges in her endeavours. One of the most serious issues she's encountered is a lack of support from government officials and other stakeholders. Many Ugandan authorities do not prioritise environmental issues, making it difficult for activists like Nakate to make significant changes. Nakate has also faced criticism and resentment from those who do not believe in the need of environmental protection.

3.4.3.1 Lessons

One valuable lesson is Nakate's perseverance to continue working despite the challenges she faces. Despite setbacks and frustrations, Nakate remains committed to her mission and continues to promote environmental protection. Nakate's life shows the importance of community and collaboration. She has been able to increase her voice and impact by working with other activists and individuals who share her beliefs.

Nakate teaches the importance of knowing and promoting awareness. Her capacity to disseminate information about environmental issues and sustainable practices has motivated others to take action and become more environmentally conscious.

Finally, she shows why it is important for environmental protection advocates to easily adapt. As an activist, she has had to be adaptive and open to new ideas and techniques to address Uganda's ever-changing environmental challenges which have helped to continue to be relevant in her advocacy.

Fig. 3.7 Image of Vanessa Nakate. Courtesy of Nakate and Credit of Laura Cook (2025)

3.4.4 Desmond D'Sa

Desmond D'Sa is a well-known South African environmental activist who has devoted his whole life advocating for environmental conservation and sustainability [44]. D'Sa was born and raised in Durban, South Africa, and has been actively involved in environmental campaigns and movements for over 20 years. He is the coordinator of the South Durban neighbourhood Environmental Alliance (SDCEA), an organisation dedicated to eliminating environmental injustices and promoting sustainable development in the neighbourhood.

D'Sa's environmental beliefs are short and straightforward. He believes that individuals and communities must assume responsibility for their environmental impact and seek to adopt more sustainable lifestyles. D'Sa advocates for natural resource conservation, pollution reduction, and the growth of renewable energy. He emphasises the need of community engagement and empowerment in environmental decision-making processes [44] (Fig. 3.8).

Despite his dedication to environmental action, Desmond D'Sa faces several challenges in his job. One of the most serious issues he faces is resistance from powerful firms and government officials who prioritise economic gains above environmental protection. D'Sa also encounters criticism and apathy from

Fig. 3.8 Image of Desmond D'Sa. Courtesy of Desmond D'Sa (2025)

certain local residents who do not fully understand the significance of environmental issues [44]. Furthermore, D'Sa often fails to get resources for his organisation, making it difficult to continue his lobbying efforts [44].

3.4.4.1 Lessons

D'Sa's activism offers key lessons including on the importance of community empowerment. D'Sa's work illustrates the power of grassroots activism and community engagement to bring positive change. He emphasises how individuals can have a significant impact on environmental issues by empowering local communities to advocate for their environmental rights, despite challenges.

He is an example of the belief that persistence pays off. Despite several challenges and setbacks, D'Sa remains committed to his mission and advocates for environmental justice. His perseverance shows that change is possible, even in the face of adversity.

His life is also a demonstration of the importance of collaboration. D'Sa often works with other organisations, activists, and stakeholders to expand his message and impact. D'Sa highlights the need of collaboration in dealing with complex environmental issues by collaborating to reach a common goal.

Lastly, his work shows that education is important in the fight for environmental protection and sustainability. D'Sa's advocacy efforts include raising awareness of environmental issues and pushing for sustainable lifestyles. He believes that education is crucial for inspiring individuals to act and make informed choices that benefit the environment

3.5 Conclusion

Traditional media outlets like newspapers, television, and community radios have been instrumental in raising public awareness about environmental issues. However, the rise of digital media has led to a decline in readership and viewership, making it more challenging to reach a large audience. Digital media, such as social media platforms, websites, and blogs, have become powerful tools for raising environmental awareness. These platforms enable individuals and organisations to share information quickly and reach a global audience. Social media campaigns can go viral, raising awareness and prompting public action. Digital media also allows for real-time engagement with the audience, allowing organisations to interact with their followers and share resources. This two-way communication leads to increased engagement and support for environmental causes. Health communication organisations must ensure creative messaging, compelling storytelling, and strategic use of media platforms. Campaigns like the Green Belt Movement, Greening Egypt, Save the Seas of Seychelles, Greening Uganda, and the Eco Warriors of Ghana have successfully mobilised people to protect the environment through their creative messaging utilising multiple communication tools including traditional and digital media. Individual environmental advocates such as Saro-Wiwa, Nakate, D'Sa, and Maathai have contributed to protection of the environment.

Recap
The media plays a significant role in shaping public opinion and raising awareness of important issues, particularly regarding environmental communication. Traditional media, such as newspapers, television, and radio, and digital media platforms, including social media and websites, both contribute to how individuals understand and relate to environmental concerns. This influence is crucial as it can change attitudes, inspire actions, and affect policies that lead to environmental improvement.

Questions

- What is the role of the media in environmental communication?
- What is the role of traditional media in environmental communication?
- What is the role of digital media in environmental communication?
- Mention at least two successful environmental communication initiatives.
- Mention three examples of individual environmental advocates in Africa.

References

1. Zhang N, Skoric MM. Media use and environmental engagement: examining differential gains from news media and social media. Int J Commun. 2018;12:380–403.
2. Mutsvairo B. Why social media activists face an uphill struggle in Zimbabwe; 2021. http://democracyinafrica.org/social-media-activists-face-an-uphill-struggle-in-zimbabwe/.
3. Wunderlich S, Gatto KA. Consumer perception of genetically modified organisms and sources of information. Adv Nutr. 2015;6(6):842–51. https://doi.org/10.3945/an.115.008870.
4. Wood TS, Baldwin S. Fuelwood and charcoal use in developing countries. Annu Rev Energy. 1985;10(1):407–29.
5. Sodeinde S. Africa's top five environmental blogs. Nat Environ. 2018; https://www.dw.com/en/africas-top-five-environmental-blogs/a-42806118
6. Kent ML, Li C. Toward a normative social media theory for public relations. P R Rev. 2020;46(1):101857. https://doi.org/10.1016/j.pubrev.2019.101857.
7. Schäfer MS, Schlichting I. Media representations of climate change: a meta-analysis of the research field. In: Media research on climate change. Routledge; 2018. p. 14–32.
8. Rahim M, Jalaladeen J. The role of social media on environmental awareness of undergraduate students in University of Sulaimani in Iraq. JALHSS. 2016;10: 218–31.
9. Kaplan AM, Haenlein M. Users of the world, unite! The challenges and opportunities of Social Media. Bus Horiz. 2010;53(1):59–68.
10. Jones JM. In the U.S., many environmental issues at 20-year low concern: worry about all eight measures tested is down from last year 2010; 2010. http://www.gallup.com/poll/126716/environmental-issues-year-low-concern.aspx.
11. Saepudin A, Mulyono D. Community education in community development. EMPOWERMENT: Jurnal Ilmiah Program Studi Pendidikan Luar Sekolah. 2019;8(1):65–73.
12. Jacobson MZ. On the causal link between carbon dioxide and air pollution mortality. Geophys Res Lett. 2008;35(3)
13. Roshandel AT, Labafi S, Robati M. Effects of social media on the environmental protection behaviour of the public (Case study: Protecting Zayandeh-rood river environment); 2016.
14. IPCC. AR6 climate change 2021: the physical science basis; 2021. https://www.ipcc.ch/report/ar6/wg1/.
15. Rheingold H. Using participatory media and public voice to encourage civic engagement. MacArthur Foundation Digital Media and Learning Initiative; 2008.
16. IFRC. What is a disaster? International Federation of Red Cross and Red Crescent Societies. 2017. https://www.ifrc.org/our-work/disasters-climate-and-crises/what-disaster#:~:text=Disasters%20are%20serious%20disruptions%20to,and%20vulnerability%20of%20a%20community.
17. Pedelty M. Environmental communication and the public sphere. Environ Commun. 2015;9(1):139–42. https://doi.org/10.1080/17524032.2014.1003440.
18. Pavelle S, Wilkinson C. Into the digital wild: utilizing Twitter, Instagram, YouTube, and Facebook for effective science and environmental communication. Front Commun. 2020;5:575122. https://doi.org/10.3389/fcomm.2020.575122.

19. Park MS. The dual role of the media in environmental communication as a public sphere and as political actors. For Sci Technol. 2013;9(1):33–8.
20. Owiny SA, Mehta K, Maretzki AN. The use of social media technologies to create, preserve, and disseminate indigenous knowledge and skills to communities in East Africa. Int J Commons. 2014;8:14.
21. Harris US. Engaging communities in environmental communication. Pac J Rev. 2017,23(1):65–79.
22. Howarth C, Lane M, Morse-Jones S, Brooks K, Viner D. The 'co' in co-production of climate action: challenging boundaries within and between science, policy and practice. Glob Environ Chang. 2022;72:102445.
23. Hillstrom K, Hillstrom LC. The Worlds environments. A continental overview of environmental issues. Bloomsbury Publishing; 2003. p. 207–8.
24. Harrison TM, Barthel B. Wielding new media in Web 2.0: exploring the history of engagement with the collaborative construction of media products. New Media Soc. 2009;11(1–2):155–78.
25. Harris US. Participatory media in environmental communication: engaging communities in the periphery. Routledge; 2018.
26. Greenpeace. 10 African youth climate activists changing the face of the planet; 2021. https://www.greenpeace.org/international/story/50006/10-african-climate-youth-activistschanging-the-face-of-the-planet/.
27. Geiger N, Swim JK, Fraser J. Creating a climate for change: interventions, efficacy and public discussion about climate change. J Environ Psychol. 2017;51:104–16. https://doi.org/10.1016/j.jenvp.2017.03.010.
28. Flor AG. Environmental communication: principles, approaches, strategies of communication applied to environmental management. UP Open University; 2004.
29. Fisher S, Bellinger DC, Cropper ML, et al. Air pollution and development in Africa: impacts on health, the economy, and human capital. Lancet Planet Health. 2021;5(10):e681–8. https://doi.org/10.1016/S2542-5196(21)00201-1.
30. Fernandez M, Piccolo LS, Maynard D, Wippoo M, Meili C, Alani H. Talking climate change via social media: communication, engagement and behaviour. In: Proceedings of the 8th ACM conference on web science; 2016. p. 85–94.
31. EPA. What is environmental education? United States Environmental Protection Agency; 2021. https://www.epa.gov/education/what-environmental-education.
32. Du Y, Wang X, Brombal D, Moriggi A, Sharpley A, Pang S. Changes in environmental awareness and its connection to local environmental management in water conservation zones: the case of Beijing, China. Sustainability. 2018;10(6):2087.
33. Cox R. Nature's "crisis disciplines": does environmental communication have an ethical duty? Environ Commun. 2007;1(1):5–20.
34. Cox R. Environmental communication and the public sphere. 3rd ed. Sage Publications; 2013.
35. Collins JM. Temperature variability over Africa. J Clim. 2011;24(14):3649–66. https://doi.org/10.1175/2011JCLI3753.1.
36. Chung CH, Chiu DK, Ho KK, Au CH. Applying social media to environmental education: is it more impactful than traditional media? Inf Discov Deliv. 2020;48(4):255–66.
37. Christensen M, Nilsson AE. Media, communication, and the environment in precarious times. J Commun. 2018;68(2):267–77.

38. Centre of Environmental Information Studies. A new model of environmental communication for Europe: from consumption to use of information: executive summary Office for Official Publications of the European Communities; 1999.
39. Facebook. Liking and reacting; 2016. https://www.facebook.com/help/452446998120360/.
40. Lester L. Three challenges for environmental communication research. Environ Commun. 2015;9(3):392–7.
41. León B, Negredo S, Erviti MC. Social engagement with climate change: principles for effective visual representation on social media. Clim Pol. 2022;22(8):976–92.
42. Lindenfeld LA, Hall DM, McGreavy B, Silka L, Hart D. Creating a place for environmental communication research in sustainability science. Environ Commun. 2012;6(1):23–43.
43. Nakate V. National Press Club; 2022. https://www.press.org/events/npc-newsmaker-vanessa-nakate-climate-activist-and-un-goodwill-ambassador. Accessed 6 June 2025.
44. The South Durban Community Environmental Alliance (SDCEA) Community News; 2024. https://sdcea.co.za/publication/. Accessed 6 June 2025.

CHAPTER 4

Role of the Traditional African Communication Methods in Environmental Communication

Introduction
For Africa to preserve and sustain its rich cultural heritage, the enduring efficacy of the traditional communication methods remain a powerful, often underestiamated asset. For millennia, African societies have depended on oral tradition and folklore as their primary means of passing along knowledge, history, and tales from one generation to the next. Even with the advancement of modern communication technologies, oral tradition and folklore remain vital living organisms of the African cultures. These traditional modes of communication offer a unique and efficient way to foster interpersonal relationships, preserve culture, and foster a sense of community. Oral tradition and folklore can be useful tools for environmental communication capable of disseminating knowledge on the natural world, fostering profound behavioural change and reinforcing indigenous environmental knowledge systems. This chapter focuses on the potential contribution of traditional African communication methods to environmental communication to preserve and sustain the natural world. This chapter begins with a discussion on the potential of oral tradition in environmental communication, then moves on to the potential of folklore in sustainable environmental communication.

4.1 Oral Traditions

In the age of technology and quick messaging, it is easy to underestimate the significance and power of oral traditions in communicating important messages, particularly when those messages deal with the environment. The

© The Author(s), under exclusive license to Springer Nature Switzerland AG 2025

G. Mutinta, *An African Perspective on Environmental Communication*,
https://doi.org/10.1007/978-3-032-05052-6_4

majority of information and ideas regarding natural Africa are handed down verbally, from generation to generation, through songs, stories, and other oral communication methods.

4.1.1 Storytelling

In Africa, storytelling has long been seen as an effective and essential means of communication. Stories have served as a means of transferring knowledge, traditions, and cultural beliefs from one generation to the next [1]. Storytelling is a vital communication tool, especially when it comes to raising awareness of environmental issues.

Stories have the power to enthral and engage audiences in ways that other forms of communication just cannot. Storytellers can engage audiences and elicit empathy for environmental challenges by weaving together compelling narratives, which is not always possible with just figures and data. When people are emotionally invested in a story, they are more likely to take action to protect the environment.

Stories can be used to educate people about the complexity of environmental issues. By crafting stories that describe the origins and consequences of issues like pollution, deforestation, and climate change, storytellers can raise awareness among the public about the need of preserving the environment. Stories can help people learn more about the environment and how their actions impact the planet.

In addition, storytelling has the power to inspire and encourage people to take action to protect the environment. By sharing the stories of individuals and groups that have had a positive impact on the environment, storytellers can show that change is possible and that everyone has a role to play in preserving the natural world. Success stories can inspire others by serving as positive role models and pushing people to make improvements in their own communities [2] (Fig. 4.1).

Stories have the ability to challenge and change perceptions of the world [2]. By telling tales that debunk popular misconceptions and stereotypes about subjects like sustainability and conservation, storytellers have the capacity to alter listeners' attitudes about the environment. By portraying the environment in a good and impactful light, storytelling has the potential to inspire individuals to see themselves as stewards of the earth instead of as passive spectators.

Using storytelling, different groups of people can collaborate and build partnerships. By telling tales that highlight the connection of all living beings and the environment, storytellers can promote a sense of unity and collective responsibility for protecting the environment. Through storytelling, environmental communication can bring people together, close gaps, and create a sense of purpose in the quest of a sustainable future [2].

4.1.2 Legends

In many African communities, the transmission of cultural values, practices, and knowledge from one generation to the next has relied heavily on legends [3].

Fig. 4.1 Storytelling. Artificial Intelligence Generated (2025)

However, in the modern world, with environmental issues becoming more urgent, legends can also be a powerful tool for environmental communication. For example, legends can be used to convey environmental messages by highlighting the relationship between humans and the environment. African stories often include animals, plants, and other natural elements as the primary characters, emphasising the importance of living in harmony with the environment. By telling these stories, organisations can inspire people to cherish and protect the environment and to realise that their actions have an immediate impact on the status of the planet.

Legends can also be used to raise awareness of certain environmental issues and challenges. By incorporating these concepts into well-known tales, campaigns can impart to people the importance of taking action to address problems like pollution, climate change, and conservation. For example, a legend about a mythical animal suffering due to a polluted river might serve as a powerful allegory for the real consequences of water pollution and inspire people to consider how their own actions could be harming the environment.

In addition, legends can be used to teach environmental care and accountability. By telling legends of ancestors who loved and protected nature, legends can remind people of their duty to protect the environment. People can be motivated by these tales to adopt eco-friendly practices and support conservation efforts as they become aware of their duty to preserve the environment for future generations.

Legends can inspire hope and motivate action to address environmental issues. By sharing stories of communities coming together to overcome adversity and protect their natural surroundings, campaigns can instil in people a sense of hope and agency. These stories can encourage people to take positive action to address environmental issues in their own lives because they make them aware of their place in a larger history of environmental care and resilience [4].

4.1.3 Myths

Myths have long been used in Africa to disseminate important information to the general public [5]. Myths, which usually provide moral lessons and an explanation of the natural world, have a tremendous effect on African culture. By incorporating environmental themes into mythology, communities can effectively communicate the need of preserving the environment in a way that is compatible with their cultural values and practices.

Myths emphasise the relationship between nature and humans, making them an effective tool for environmental communication. Many African myths depict spirits and deities as being in the natural world, emphasising the value of living in harmony with the environment. By incorporating these concepts into environmental messaging, communities can be better able to understand how their actions affect the environment and strive to protect it for coming generations.

Myths can also be used in the environmental communication field to support conservation efforts through storytelling. Mythological characters often have to overcome challenges or make tough choices in order to rescue the environment. By sharing these tales with others in the community, people can impart valuable knowledge about the need of conservation and the consequences of failing to protect the environment.

Communities can be educated via myths about the importance of biodiversity and the necessity to preserve endangered species. Numerous African myths feature prominent animal characters, endowing them with symbolic meaning and imparting lessons on the need of reverencing and safeguarding wildlife [6]. Through the use of these themes in environmental communication, communities can work to protect the many ecosystems that make up the natural world and develop a greater respect for it.

In addition, myths are a powerful tool for environmental communication because they can inspire communities to adopt sustainable practices and take up issues. Stories usually include heroes who overcome great odds and accomplish incredible feats, setting an example for others to follow. By incorporating stories of environmental activism and conservation efforts into mythology, communities can be inspired to take action against environmental degradation and work towards a more sustainable future [7].

4.1.4 *Folk Tales*

A significant part of African culture, along with other traditional modes of communication, are folktales [7]. In addition to being used for entertainment, these stories are passed down from generation to generation as a means of transferring knowledge and wisdom. Studies show that folktales have come to be recognised as powerful tools for environmental communication [3].

Folktales can be used to highlight the relationship between people and the natural world. In African folktales, animals are usually the main characters and are usually shown as intelligent, powerful beings. Folktales, which depict the coexistence of humans and animals, can educate individuals about their integral role within the ecosystem and the immediate consequences of their actions on the environment. This can inspire individuals to adopt more environmentally friendly habits and foster a sense of responsibility for the environment.

Besides, folktales have the ability to serve as alerts about the harmful consequences of environmental degradation. Many traditional African stories contain themes of greed, disdain for the natural world, and the need of living in harmony with the environment. Communities that tell these stories and emphasise their moral lessons might raise public awareness of the dangers posed by pollution, deforestation, and other environmental issues. This can inspire people to take action to protect the environment and foster a respect for the natural world (Fig. 4.2).

Moreover, people can be inspired to take proactive measures to safeguard the environment by folktales. African folktales usually feature heroic, resourceful, and fearless characters who overcome insurmountable challenges. Communities can encourage others to get involved in environmental conservation by sharing the experiences of regular people who have had a positive impact on the environment. People could feel more empowered therefore to advocate for sustainability and environmental protection in their local areas.

Folktales can be used to impart traditional ecological knowledge and sustainable practices to the public. Since many African communities have a deep knowledge of their natural surroundings, they have developed long-standing customs for living in harmony with the environment. By incorporating this traditional knowledge into folktales, communities can disseminate crucial information about resource management, biodiversity preservation, and sustainable agriculture. Indigenous knowledge and traditions can be preserved in this way to sustain ecological balance and resilience in the face of environmental challenges.

Fig. 4.2 Folk tale imaginary image. Courtesy of Wikimedia Common (2024)

4.2 Folklore

Folklore has an important—though occasionally overlooked—role in environmental communication [8]. It can be used to raise awareness among the general public, motivate action, and support conservation. Folklore encompasses the traditional beliefs, customs, and stories passed down orally from generation to generation within a particular community. It is a reflection of a culture's heritage and identity, shaping the values and practices of its people. Folklore includes dances, legends, riddles, fairy tales, proverbs, songs, and rituals that are shared among members of a society and have the power to communicate important environmental messages [9–11].

4.2.1 Proverbs

Proverbs are a useful tool for disseminating important knowledge about environmental preservation and sustainability [12]. African proverbs are a powerful technique to communicate important concepts in a way that is understandable to others and has a strong cultural effect. Proverbs that have meaning and are well-known in the community can be used to convey efficiently and relatably environmental themes [13]. To demonstrate the need of sustained environmental conservation initiatives and the need for perseverance and patience in environmental preservation, is reflected for example in the proverb "A tree does not grow in a day."

As a result of their complexity, environmental issues can be difficult for people to understand, especially in rural regions where access to information can be limited. Proverbs provide a plain, intelligible explanation of these concepts [14]. As the proverb "You cannot sell the cow and drink the milk" demonstrates, people cannot utilise the environment without consequences.

Proverbs have the power to inspire action and promote the formation of positive new routines. Proverbs that emphasise the importance of sustainable activities, such as "A clean bird builds a clean nest," can encourage people to adopt eco-friendly practices in their daily lives. Proverbs can serve as a useful reminder of the interdependence between humans and the natural world and the need to preserve the environment for future generations.

The collection of proverbs is a shared cultural history that fosters a sense of community and unifies individuals. If proverbs are used to communicate environmental concerns, communities can come together and work towards common goals and solutions. Proverbs can serve as a catalyst for collective action by serving as a reminder of people's shared responsibility to protect the environment.

4.2.2 Riddles

Riddles play a major role in African culture's information sharing, knowledge transfer, and community development [15]. They are both a type of entertainment and an efficient instrument for environmental communication.

African communities have long used riddles to convey important information about the environment, wildlife, and available resources [16]. They serve to pique people's curiosity, challenge preconceived notions about the world, and promote a deeper understanding of environmental issues.

Riddles can be used to educate people about biodiversity and conservation. By figuring out riddles regarding different plant and animal species, people can get more knowledge about the need of preserving biodiversity and protecting endangered species. As a result, people could become more accountable and motivated to take action to save the environment for future generations.

Riddles can be used to highlight the benefits of acting in an ecologically responsible manner as well as the negative effects of environmental degradation. If riddles, for example, concerning climate change are included, people can become more aware of how their actions impact the environment. This can encourage people to adopt environmentally friendly habits and spark discussions about the need for long-term solutions.

Promoting indigenous practices and traditional knowledge about environmental management can be accomplished with the use of riddles. Communities can share riddles that convey ancient wisdom about coexisting peacefully with nature, thus preserving their cultural heritage and passing on valuable knowledge to next generations. This can inspire individuals to uphold enduring norms and foster a sense of cultural pride.

Riddles can be an excellent tool for coming up with original solutions to environmental issues [17, 18]. Through the process of solving riddles that require ingenuity and creative thinking, people can develop problem-solving skills and a resilient mindset during environmental calamities. This can allow individuals the self-assurance to take on challenging environmental issues head-on and find lasting solutions.

4.2.3 Chants

Africans have been using chants for many years. Chants are rhythmic vocalisations that are occasionally accompanied by music or dancing [19]. Chants can be a powerful tool for environmental communication because they can convey complex thoughts and concepts in a memorable and engrossing way. For example, by including topics such as the importance of conservation, sustainability, and the preservation of natural resources into their chants, communities can inform and educate people about the pressing environmental concerns that impact their community. Chants have a way of capturing listeners' attention and sticking in their memory, making them an effective tool for bringing attention to issues and inspiring people to take action.

Chants are a helpful technique for encouraging individuals in a group to conserve the environment. Communities can employ chants to motivate people to work together towards a common goal, thereby utilising the power of collective action to successfully tackle environmental concerns. Chants have a way of bringing people together and fostering a sense of purpose, which makes them a useful tool for encouraging communities to work towards a sustainable future.

Besides, chants can be used to respect and honour the natural world [20]. By including phrases expressing gratitude and admiration for the diversity and beauty of the natural world into their chants, communities can inspire individuals to treat the environment with care and respect. Chants, since they evoke a sense of surprise and awe, are an effective means of fostering a sense of environmental stewardship and responsibility.

It is possible to pass on ecological wisdom and information from one generation to the next through chants [20]. By including teachings about the need of living in harmony with nature, enjoying the land, and taking care of the earth into their chants, communities can ensure that crucial environmental information is preserved and passed on to future generations. Chants have the power to uphold and spread cultural traditions and beliefs, which makes them a valuable tool for preserving and promoting environmental wisdom and practices [21].

4.2.4 *Songs*

African communities have traditionally used songs to communicate messages to an individual or communities [22]. Songs are a powerful tool that played a crucial role in transferring knowledge, cultural values, and historical details from one generation to the next.

Songs can captivate listeners and convey concepts in an engaging and lasting fashion [23]. By including environmental themes into their songs, songwriters can raise awareness of significant environmental issues. Through enduring melodies and powerful lyrics, songs can inspire listeners to take action by teaching them the importance of environmental conservation.

Songs can also be used to promote conservation efforts and highlight how important it is to preserve biodiversity and natural regions. Musicians can use their platform to advocate for wildlife conservation, habitat restoration, and ecologically responsible agricultural methods. Singing about the wonders of nature and the need to preserve it, artists can encourage their audience to support conservation initiatives and adopt environmentally conscious decisions in their everyday lives (Fig. 4.3).

Many African communities have used traditional songs as a way to spread knowledge about sustainable practices [25], such as land stewardship, water management, and resource conservation [26]. By incorporating these historical lessons into current songs, musicians can encourage communities to adopt eco-friendly practices and support sustainable development. Songs have the potential to be a powerful tool for advancing sustainable development policies that find a balance between economic growth and environmental protection.

Fig. 4.3 Singing. SDCEA (2024) [24]

Music has the power to unite people and motivate groups to strive towards a common objective. Songs can be used to inspire people to support campaigns, protests, and advocacy activities in the context of environmental communication. It is possible for artists to compose music that inspires listeners to unite in the fight against environmental degradation and climate change. Through the collective power of music, communities can come together to address pressing environmental challenges and work towards a more sustainable future.

4.2.5 Poetry

Poetry has always been a useful tool for communication in African communities [27]. Therefore, poems can be a powerful tool for promoting environmental awareness. For example, poetry can be used to spread awareness of environmental issues by raising public awareness of these issues. Poets have the power and beauty of language to educate people about the need of environmental conservation. Poets who use vivid imagery and intense emotion in their poems might help listeners comprehend the problems facing the environment and develop a sense of connection with it. Poetry can inspire readers to act individually or collectively to protect the environment.

Teaching the public about how human activity affects the environment is another way poetry can be used to communicate environmental issues. Poets can use language to communicate complex environmental concepts in an engaging and clear manner. Poets can help audiences see the need of sustainability and conservation by breaking down difficult environmental subjects into manageable chunks. Poems can highlight the interconnectedness between all living things and the consequences of environmental degradation.

Through poetry, poets can challenge the status quo and demand change in response to environmental challenges. Since poets highlight the seriousness of the issue and the need for immediate action, communities might be motivated to demand changes in laws and resource allocation to protect the environment. Poems have the power to inspire people to fight for environmental justice and to work for a better future.

In addition, poetry can be used to praise nature's wonders and beauty. Poets can remind readers of the value of the environment by illustrating the amazing landscapes, diverse ecosystems, and fascinating animals that make up our world. Poetry cultivates a sense of reverence and thankfulness for the Earth, which might encourage people to protect it for future generations. Poets highlight the wonders of nature to inspire others to care for the environment.

4.2.6 Oral History

Throughout Africa, stories, customs, and historical accounts have been transmitted from one generation to the next via oral history, in which knowledge is transmitted verbally instead of via written texts [28, 29]. This is one of the most efficient forms of communication in African communities. Since oral history allows for the efficient and culturally acceptable sharing of environmental knowledge and experiences, it can be a prominent tool for environmental communication in recent years.

Complex ecological knowledge, including information on flora, wildlife, and ecosystems that has been passed down orally for many generations, is possessed by many African indigenous tribes. By recording and sharing this knowledge through oral history, campaigns can help to preserve significant insights into methods of sustainable resource management, biodiversity conservation, and climate change adaptation [30].

Oral history can be used to incorporate community members in discussions about environmental issues that directly impact them. By sharing personal environmental stories and experiences, communities can come together to generate ideas, set priorities, and mobilise resources for sustainable development efforts.

Moreover, oral history is another useful tool for challenging popular environmental narratives. The marginalisation of indigenous knowledge systems in favour of Western concepts and scientific knowledge has occurred often, leading to misunderstandings and conflicts over natural resources such as land and water [31, 32]. By utilising oral history, environmental communication can provide voice to marginalised communities, influence the way environmental issues are addressed, and promote more cooperation and understanding.

In addition, oral history can inspire activism and take action for environmental conservation. By sharing oral stories of stewardship, adaptation, and resilience, environmental communication can encourage individuals and groups to take action to preserve and protect their natural environments for coming generations. People can increase their sense of connectivity and responsibility to the environment by telling tales, which can help to foster a culture of environmental stewardship and care.

4.2.7 *Proclamations and Announcements*

In Africa, proclamations and announcements, which are widely used, can be useful tools for environmental communication [33]. For example, proclamations and announcements can be used to instil in local populations the importance of environmental preservation. Through traditional communication channels like village meetings or community gatherings, leaders and elders can spread knowledge about sustainable practices, conservation activities, and the consequences of environmental degradation on day-to-day living. By raising awareness, this form of communication can motivate community members to take action to protect their natural environment.

Proclamations and announcements have the power to spur action within the group. Leaders who openly voice their concerns about the environment and inspire community people to come together for a common goal can build support and foster a sense of unity and solidarity. These announcements can be used by communities to organise tree-planting events, clean-up days, and other environmental conservation initiatives.

By making pronouncements and announcements, communities can enforce environmental laws and policies. Through public announcements, leaders can set clear guidelines and laws that ensure the sustainable use of natural resources and environmental protection, so creating norms and standards. Through these alerts, the community can be made aware of the consequences of violating these guidelines and the prohibition on activities that harm the environment, such as poaching, illegal logging, and littering.

Proclamations and announcements can be used to recognise and celebrate environmental achievements [34, 35]. Publicly honouring people or groups that have made noteworthy contributions to environmental preservation can inspire others to pursue similar goals and serve as a role model for others. By making these announcements, communities can honour conservation heroes, showcase success stories, and cultivate a sense of pride and commitment in protecting their environment for future generations.

4.2.8 Tribal Meetings

Tribal meetings can be used in environmental communication. These gatherings help the community with communication, decision-making, information sharing, and problem resolution [36, 37]. Nowadays, it is thought that tribal meetings are a good way to talk about the environment. For example, tribal meetings give the local populace a platform to discuss environmental issues that directly affect their daily life. By gathering in a public area, people can share knowledge, worries, and first-hand accounts on pollution, deforestation, climate change, and environmental degradation. By speaking and debating these pressing environmental issues, community members can come up with solutions and strategies (Fig. 4.4).

Tribal meetings can be utilised to provide vital information and raise awareness of methods for protecting the environment. Local government agencies and environmental experts can utilise these occasions to educate the public about the need of preserving biodiversity, recycling, composting, conserving water, and protecting natural resources. By imparting information to community members, tribal meetings have the potential to instil a sense of duty and care for their local environment [38].

Tribal gatherings can provide a venue for promoting environmental initiatives and advocating for policy changes at the local, provincial, and national levels. By speaking with one voice, tribal groups can utilise their combined influence to support the establishment of protected areas, tighten environmental regulations, and advance sustainable development projects. When it comes to making choices, tribal meetings have the capacity to influence lawmakers and decision-makers to prioritise environmental sustainability and conservation.

Fig. 4.4 Tribal meetings. Courtesy of Wikimedia Common (2022)

Moreover, tribal meetings have the ability to motivate community-led conservation initiatives. These gatherings have the capacity to inspire individuals to take initiative and implement locally based environmental solutions by fostering a sense of shared ownership and responsibility among attendees. Through waste management initiatives and reforestation projects, tribal meetings can mobilise resources—including money, people, and experience—within the community to address environmental issues and bring about positive changes.

In addition, tribal meetings can foster a sense of cultural pride and traditional knowledge about environmental protection methods. Many African indigenous tribes have been sustained for decades on the land, wildlife, and natural resources that are intrinsic to their identity [38–42]. By incorporating traditional beliefs, practices, and rituals into environmental communication programmes, tribal gatherings can serve to perpetuate the cultural inheritance and values that promote sustainable living and harmonious coexistence with nature.

4.2.9 Elders and Wisdom

In many African communities, the transmission of knowledge and practices from one generation to the next depends heavily on elders and wisdom keepers [38, 40]. These distinguished individuals are recognised as the custodians of wisdom, skill, and cultural heritage. Because of this, they could work well as tools for environmental communication. For example, elders and knowledge keepers can communicate about the environment via storytelling. African communities have traditionally used oral storytelling as a means of passing along knowledge and ideas. Elders can share tales about the land, animals, and natural resources with the community to help them understand the need of preserving the environment for future generations. A sense of one's responsibility and stewardship for the environment can be fostered by these stories.

Elders and knowledge keepers can also use rituals and ceremonies as a means of communicating with the environment. African cultures have several rituals and traditions that are related to the natural world [40]. By including lessons about sustainability and conservation and involving elders in these activities, communities can get a greater understanding of the connection of humans and the environment. Elders can also offer blessings and prayers that emphasise the value of preserving and safeguarding the environment (Fig. 4.5).

Furthermore, elders and knowledge keepers can serve as community role models for sustainable living. By setting a good example, these well-respected individuals can inspire others to adopt eco-friendly practices and lifestyles. Senior citizens can serve as role models for sustainable living by engaging in ethical resource harvesting, traditional farming methods, and energy and water conservation.

Discussions in the community and environmental decision-making processes can involve elders and wisdom keepers. Their experience and knowledge

Fig. 4.5 Elders and wisdom. Artificial Intelligence Generated (2024)

can provide perceptive comments and points of view that younger members of the community cannot have considered. The richness of experience and knowledge that comes from elders in discussions on land use, conservation efforts, and sustainable development projects can benefit communities.

4.2.10 Rituals

African traditional communication methods have always relied on rituals as a reliable source of information [43–45]. Rituals can be used in the sector of environmental communication, where stressing the need of sustainable practices and conservation efforts is crucial.

Rituals hold significant cultural and spiritual value in African communities. By incorporating environmental teachings into their rituals and celebrations, communities can establish profound relationships with the natural world. For example, celebrations of the changing of the seasons or rituals honouring local deities can serve as a platform for discussing the importance of preserving the environment and living in harmony with nature.

Rituals usually include a common experience that binds communities together [44]. Rituals might be a means of bringing environmental concepts to a wider audience and starting dialogues that lead to collective action. A custom that involves doing things like planting trees or cleaning up a nearby river might encourage a sense of community and shared accountability for environmental preservation (Fig. 4.6).

Fig. 4.6 Rituals. Courtesy of Wikimedia Common (2022)

Moreover, rituals have the power to reach people's emotions and beliefs in a way that traditional, Western communication methods cannot be able to. Communities can be made more receptive to the importance of conservation efforts and sustainable practices by incorporating environmental information into intensely emotive or ritualistic acts. For example, a tradition of planting trees in honour of ancestors might emphasise the relationship between the past, present, and future generations and the need to protect the environment for them.

Rituals have a lasting impact on individuals and groups, progressively changing attitudes and actions [46–50]. Rituals that are passed down through the generations can incorporate environmental themes to foster long-lasting environmental stewardship and respect among communities. For example, rituals honouring the seasons or agricultural cycles can be a useful reminder of how crucial it is to live in harmony with nature and adapt to its cycles.

4.2.11 Dances

One of the oldest and most widely used forms of traditional African communication is dancing [26, 44]. People have used dance to celebrate cultural traditions, tell stories, and deliver important messages for ages. Therefore, dancing can also be used as an environmental communication tool. For example, dances can be used to raise awareness of environmental issues. The detrimental consequences of pollution, deforestation, climate change, and other environmental challenges can be demonstrated via dances. Seeing these dances can educate viewers on the ways that environmental degradation affects both local and global cultures.

Dances can also be used to promote conservation and sustainable behaviours. Dances that emphasise the importance and beauty of wildlife, plants, and natural habitats can be made. Communities can be inspired to protect their natural habitats and advance sustainable living by using dance to highlight the importance of conservation (Fig. 4.7).

People can be inspired and encouraged to take action for the environment by dancing. Choreographed dances can be used to draw attention to community-based initiatives, push for legislative changes, and inspire people to take up environmental causes. Dancing can be used as an advocacy tool to help communities come together to safeguard the environment for future generations.

In addition, dances can also be used to promote environmental stewardship and fortify cultural relationships. By including traditional dances in environmental communication campaigns, communities can honour their cultural heritage and increase public awareness of environmental issues at the same time. Dancing can bring communities together to share knowledge and insights, learn from one another, and work towards the common goal of protecting the environment.

Fig. 4.7 Dance. SDCEA (2024) [24]

4.3 Conclusion

Traditional African communication methods, such as oral tradition and folklore, have the potential to significantly contribute to sustainable environmental communication. Oral traditions, passed down through generations, have the unique ability to engage listeners and convey important messages in a compelling and memorable way. By incorporating environmental themes into these traditions, communities can effectively spread awareness about the importance of environmental conservation and sustainability. Folklore plays a significant role in shaping cultural attitudes and beliefs, and integrating environmental messages into traditional stories and myths can influence behaviour and inspire action. Oral history and folklore methods are deeply rooted in the cultural fabric of African societies, making them highly effective tools for engaging communities in conversations about environmental issues. By tapping into these rich oral and folklore traditions, environmental communicators can connect with people in a meaningful way and inspire them to take action. Additionally, traditional African communication methods can bridge the gap between indigenous knowledge and modern environmental science, resulting in greater understanding and collaboration between different knowledge systems, ultimately leading to more effective sustainable environmental communication conservation and sustainability efforts.

Questions

- What is your understanding of the concept of African oral tradition and folklore?
- What are some of the African oral tradition and folklore?
- What is the potential of African oral tradition and folklore in environmental communication?

Recap

African traditional communication tools have been integral to the continent's culture, history, and social organisation. These tools, which include storytelling, chants, songs, proverbs, dances, riddles, drumming, and communal gatherings, possess unique qualities that can be effectively utilised in sustainable environmental communication. As the world faces pressing environmental issues such as climate change, deforestation, and biodiversity loss, leveraging these traditional methods can enhance awareness and foster community engagement in sustainable practices.

References

1. Cramer CM. The framing of climate change in three daily newspapers in the Western Cape Province of South Africa. Doctoral dissertation. Stellenbosch: University of Stellenbosch; 2024. https://scholar.sun.ac.za:443/handle/10019.1/2204.
2. Death C. Environmental movements, climate change, and consumption in South Africa. J South Afr Stud. 2014;40(6):1215–34. https://doi.org/10.1080/03057070.2014.964494.
3. Dlamini JS. Safari nation: a social history of the Kruger National Park. Ohio University Press; 2020.
4. Evans HC. Re-articulating media re/presentations of climate change discourse (s) in South Africa: climate change politics in the Global South. Doctoral dissertation. University of KwaZulu-Natal, Howard College; 2019.
5. Evans HC, Dyll L, Teer-Tomaselli R. Communicating climate change: theories and perspectives. Handbook of climate change communication: Vol. 1: theory of climate change communication. Springer International Publishing; 2018. p. 107–22.
6. Leal Filho W, Manolas E, Azul AM, Azeiteiro UM, McGhie H. Handbook of climate change communication: Vol. 2. Practice of climate change communication. Springer; 2018.
7. Francis D, Valodia I, Webster E. Inequality studies from the global south. London: Routledge; 2020. https://doi.org/10.4324/9780429282447.

8. Francis D, Webster E. Poverty and inequality in South Africa: critical reflections. Dev South Afr. 2019;36(6):788–802. https://doi.org/10.1080/0376835X.2019.1666703.
9. Ghosh A. Can the non-human speak? Other beings in myth, literature and ethnography. Yale NUS College; 2019. https://www.yale-nus.edu.sg/events/amitav-ghosh-can-the-non-human-speak/
10. Ghosh A. The great derangement: climate change and the unthinkable. UK: Penguin; 2018. https://www.bibliovault.org/BV.landing.epl?ISBN=9780226323176
11. Grebowicz M. Glacial time and lonely crowds: the social effects of climate change as internet spectacle. EH. 2014;5(1):1–11. https://doi.org/10.1215/22011919-3615388.
12. Green L. Rockl waterl life: ecology and humanities for a Decolonial South Africa. Duke University Press; 2020.
13. Haraway D. Situated knowledges: the science question in feminism and the privilege of partial perspective 1. In: Women, science, and technology. Routledge; 2013. p. 455–72.
14. Haraway D. When species meet. University of Minnesota Press; 2008.
15. Sibande M. Dance for the earth: ecological themes in traditional African performance. J Afr Perform Stud. 2018;22(4):315–28.
16. Mwathi N. The power of poetry in raising awareness about environmental issues in Africa. J Environ Lit. 2020;18(3):56–68.
17. Odumosu S. Preserving oral environmental history in African communities. Afr Her J. 2017;14(2):75–86.
18. Tembo J. Oral history and the environment: documenting ecological knowledge in African communities. Afr Hist Rev. 2023;38(2):189–201.
19. Kamau S. Oral history and its importance in preserving environmental knowledge in Africa. Afr J Oral Trad. 2019;20(3):112–25.
20. Lütz JM, Mendy E, Totin E, Filho WL, Ayal D. Obstacles to implementing indigenous knowledge in climate change adaptation in Africa – ScienceDirect. www.sciencedirect.com. 2025. https://www.sciencedirect.com/science/article/pii/S0301479724036648.
21. Kamau S. The significance of oral history in sustainable development: perspectives from East Africa. J Sust Dev. 2016;4(1):112–25.
22. Ibikunle O. Community proclamations on environmental preservation in Africa. Afr Dev Stud. 2018;33(4):201–15.
23. Dube S. Proclamations and announcements: communication tools for environmental advocacy in Africa. Afr J Comm Stud. 2024;30(2):113–25.
24. The South Durban Community Environmental Alliance (SDCEA). Community News. 2024. https://sdcea.co.za/publication/. Accessed 6 June 2025.
25. Odhiambo B. The role of proclamations in raising awareness about environmental issues in Africa. J Commun Stud. 2017;14(5):67–79.
26. Okonkwo N. The wisdom of African elders in environmental communication. ESR. 2020;18(3):112–25.
27. Chigunta P. Tribal meetings: a platform for environmental dialogue in African communities. J Environ Comm. 2015;20(4):389–402.
28. Lungu K. Elders and wisdom: the role of traditional knowledge in African environmental communication. Int J Afr Stud. 2016;15(1):45–58.
29. Njoroge P. Tribal meetings and their significance in addressing environmental challenges in Africa. J Afr Com Dev. 2021;19(2):57–69.

30. Achebe C. Things fall apart. Anchor Books; 2015.
31. Obiero M. Elders and their wisdom in promoting sustainable practices in African communities. Afr J Sust Dev. 2022;27(1):23–36.
32. Tidjani A. Embracing elders' wisdom for sustainable environmental practices: a case study of the Fulani people in Nigeria. Environ Ethics. 2024;31(1):76–89.
33. Chipeta M. The rituals of sustainability: African understanding of ecology and the environment. Environ Conserv. 2017;44(3):235–45.
34. Nyambura A. The cultural significance of dances in communicating environmental messages in Africa. Int J Cult Stud. 2020;15(3):76–89.
35. Waweru B. The significance of rituals in honoring the environment in Africa. Afr J Rit Prac. 2016;18(4):88–101.
36. Ahmed M. The role of indigenous communication systems in environmental conservation: a case study of the Maasai Community in Kenya. J Afr Cult Stud. 2017;29(3):325–40.
37. Bello R. Harnessing the power of folklore in environmental communication: insights from West Africa. J Environ Comm. 2019;12(2):187–201.
38. Diop A. Communicating climate change through African proverbs: lessons from Senegal. Afr J Environ Sci Technol. 2018;7(4):87–98.
39. Moyo I. Integrating environmental themes into African myths and legends: a path to conservation. Conserv Biol. 2023;27(2):248–65.
40. Liam CU, Koffi DN, Bwalya C, Sènankpon T, Adrian W, Naomi T, Laura MP. African futures: a review of scenarios for Indigenous and local people and nature in Africa. Ecol Soc. 2024; Ecologyandsociety.org. https://ecologyandsociety.org/vol29/iss3/art32/
41. Adams M. Indigenizing the Anthropocene? Specifying and situating multi-species encounters. IJSSP. 2021;41(3/4):282–97. https://www.emerald.com/insight/content/doi/10.1108/ijssp-04-2019-0084/full/html
42. Ahmed S. The cultural politics of emotion. Routledge; 2013. https://www.taylorfrancis.com/books/mono/10.4324/9780203700372/cultural-politics-emotion-sara-ahmed-sara-ahmed
43. Barad K. Meeting the universe halfway: quantum physics and the entanglement of matter and meaning. Duke University Press; 2007. https://doi.org/10.2307/j.ctv12101zq.
44. Bosch T. Blogging and tweeting climate change in South Africa. Ecquid Novi: Afr J Stud. 2012;33(1):44–53. https://doi.org/10.1080/02560054.2011.636825.
45. Brenner J, Hofmeyr I, Lavery C. Water/colour/bead. IMS. 2022;2(2):131–40. https://doi.org/10.1177/263497952210871.
46. Burnett S. White belongings: race, land, and property in post-apartheid South Africa. Rowman & Littlefield; 2022.
47. Castro-Sotomayor J. Territorialidad as environmental communication. AnnCom. 2020;44(1):50–66. https://doi.org/10.1080/23808985.2019.1647443.
48. Chevallier N, Wood L. (Directors). Wild coast warriors [Video recording]; 2024. https://www.wildcoastwarriorsfilm.com.
49. Childs JR, Hicks CC. Securing the blue: political ecologies of the blue economy in Africa. J Polit Ecol. 2019;26(1):323–40. https://journals.librarypublishing.arizona.edu/jpe/article/id/2116/
50. Clark T. Ecological grief and Anthropocene horror. Am Imago. 2020;77(1):61–80. https://doi.org/10.1353/aim.2020.0003.

CHAPTER 5

Challenges and Opportunities for Environmental Communication

Introduction
There are challenges and opportunities for environmental communication in Africa when it comes to conveying the importance of sustainability and environmental protection. Despite the continent's natural splendour, ecological significance, and rich cultural diversity, there are a number of barriers to effective communication that need to be overcome. Nevertheless, if the right strategies and tactics are applied, positive change and greater understanding of Africa's environmental problems are achievable. This chapter discusses challenges to environmental communication in Africa followed by significant opportunities for environmental communication in addressing environmental issues.

5.1 Challenges to Environmental Communication

The unique challenges and potential that environmental communication in Africa presents cannot be ignored. On a continent rich in biodiversity, natural resources, and cultural heritage, it is imperative to use effective communication tactics in order to address pressing environmental issues. Even though the challenges can seem insurmountable, there are plenty of opportunities to include and motivate communities to embrace environmentally friendly practices.

5.1.1 Lack of Resources

A huge resource shortage in Africa is impeding environmental communication efforts [1]. This issue eventually makes it more difficult to properly spread information and education on environmental problems, which impedes efforts

© The Author(s), under exclusive license to Springer Nature Switzerland AG 2025

G. Mutinta, *An African Perspective on Environmental Communication*, https://doi.org/10.1007/978-3-032-05052-6_5

to protect and maintain the continent's natural resources. For example, Africa's lack of resources severely limits the infrastructure needed for effective communication. Due to inconsistent Internet connectivity, it is difficult for people in many rural African areas to access and exchange information online. Furthermore, there are fewer traditional media outlets, such as newspapers and radio stations, in rural areas. Promoting sustainable practices and educating the public about environmental issues becomes much more challenging in the absence of these communication channels.

Lack of funds makes it difficult to develop effective educational materials and campaigns on environmental issues in Africa. Research, graphic design, printing, and distribution all require funding to create engaging and instructive content [2]. Without these tools, governments and organisations struggle to develop and implement effective communication strategy. As a result, the general public can continue to lack knowledge or understanding regarding environmental problems and possible solutions.

Moreover, the lack of resources in Africa makes it difficult for governmental organisations and environmental organisations to implement outreach and engagement programmes. To raise awareness and encourage habit change, community workshops, school programmes, and public forums are essential. Nevertheless, funding for personnel, transportation, and logistical support is often not provided for these initiatives. As a result, opportunities for direct public engagement are limited, which makes it more challenging to promote action and improve sustainability.

Lack of resources in Africa result in the marginalisation of vulnerable populations in environmental communication. Sometimes fiscal constraints and systemic inequalities lead to the exclusion of women, youth, and indigenous people from communication channels and decision-making processes. Without targeted outreach and tailored messaging, these groups cannot be able to get the information and support they need to advocate for their rights and participate in environmental initiatives. As a result, the views of marginalised groups are often silenced, worsening inequality and upholding environmental injustices.

5.1.2 *Lack of Infrastructure*

In many parts of Africa, there is limited access to basic technology, such as computers, the Internet, and telephones [3]. This lack of access hinders communities' ability to acquire and share knowledge about environmental problems. Without access to technology, people cannot stay informed about conservation efforts, climate change concerns, and sustainable practices. Ultimately, this reduces the effectiveness of environmental communication in terms of educating and engaging the community.

Inadequate transportation and communication networks make it difficult for environmental messaging to reach isolated populations in isolated rural areas. Because there are no stable roads, public transportation, or postal services in these places, it is challenging to conduct outreach efforts and distribute information. This disconnection makes it difficult to effectively communicate with underprivileged groups, who are often the ones most impacted by environmental degradation and climate change.

Besides, Africa's media landscape faces several challenges, including a lack of professional training, government censorship, and insufficient funding. It is more challenging for media outlets and journalists to appropriately report environmental issues as a result of these restrictions. Without a strong media infrastructure, environmental communication projects struggle to gain momentum in the public discourse. The dearth of comprehensive and accurate information on environmental concerns makes it more difficult to raise awareness and motivate action on serious environmental matters.

The lack of public venues such as libraries, community centres, and information hubs in Africa poses an additional challenge to environmental communication activities. These spaces are required to provide training sessions, public lectures, and seminars educating the community about environmental problems and solutions. When such facilities are not accessible, it is challenging to engage with a range of people and foster meaningful conversation on environmental problems. This lack of public infrastructure limits the reach and effectiveness of environmental communication initiatives in Africa.

5.1.3 Language Barriers

The fact that so many languages are spoken in Africa, not everyone can have access to important environmental information [4]. This limits the reach of awareness campaigns and instructional materials, which increases the difficulty of enlisting the public and motivating them to care for the environment. People cannot be aware of how their actions influence the environment or understand the need of conservation efforts if they do not have access to relevant information.

When translating environmental information into other languages, misinterpretations and misunderstandings can occur. Translations that lose language subtleties and complexity might lead to misunderstandings and incorrect information. This might result in misaligned actions and inadequate communication, impeding the progress of environmental initiatives. To ensure accurate and clear communication, it is essential to narrow the language gap and foster linguistic diversity in environmental communication activities.

Language barriers can also prevent stakeholders, intercommunal groups, and organisations from engaging on environmental issues. Forming alliances and organising collaborative ventures across national boundaries is challenging when linguistic obstacles prevent efficient communication. This can make it more challenging to exchange resources, knowledge, and best practices, which can make it more difficult to implement viable long-term environmental solutions. By overcoming linguistic barriers, it is possible to encourage more unity and solidarity in the fight against environmental degradation.

In addition, language barriers can also cause some groups within society to become alienated and disadvantaged, particularly for individuals who speak minority or indigenous languages. When environmental information is mostly provided in an official or dominant language, it can exacerbate inequalities and marginalise these groups of people. In addition to being denied access to critical environmental knowledge, their participation in environmental governance and decision-making processes can also be jeopardised. Prioritising multilingual communication in the environmental movement is essential to promoting linguistic variety and advancing inclusivity.

5.1.4 *Limited Access to Information*

One of the reasons that information access in Africa is constrained is the lack of appropriate infrastructure [5]. It is difficult for residents of many remote communities to acquire vital environmental information because they do not have access to reliable Internet connectivity or electricity. This absence of infrastructure creates a digital divide that keeps some communities behind in the age of information. When people lack access to the Internet or other kinds of communication, they are less likely to adopt sustainable practices and are unable to stay informed about environmental problems.

Africa is a linguistically diverse continent where over 2000 different languages are spoken [6]. In many of these countries, environmental information is only available in colonial languages like English, French, or Portuguese, which are not widely spoken by the local public. Because of this language barrier, people are unable to access vital environmental information that is usually inaccessible in their native tongue. As a result, many communities are kept in the dark about environmental challenges and are unable to participate in relevant discussions or initiatives.

Education is crucial to enhancing environmental communication and promoting sustainable behaviour. However, many African countries find it difficult to give their inhabitants access to good education, particularly in rural areas. People who lack easy access to schools, qualified educators, and educational resources are unable to acquire the knowledge and skills necessary to understand environmental problems and take part in meaningful talks. Without a solid educational foundation, people are less likely to recognise the need of environmental conservation and to take action to protect their natural resources.

5.1.5 *Government Censorship and Media Restrictions*

Government censorship and media restrictions prevent journalists and environmental activists from covering environmental issues [7]. Many African countries impose stringent regulations on the media, suppressing any voices seen to be critical of their actions or policies [8]. This makes researching and writing on environmental issues like illegal logging or water pollution particularly difficult for journalists. This can mean that the general public is unaware of the gravity of these issues or the urgency with which they must be remedied.

When environmental issues are censored by the government, propaganda and misleading information can emerge. When governments control the flow of information, they have the ability to rewrite history to suit their own goals. This can result in skewed reporting that downplays the importance of environmental problems or even denies their existence altogether. When this occurs, the public is not informed about the true state of the environment and cannot be motivated to take action to protect it.

Besides, people can find it more difficult to engage in environmental decision-making when there are restrictions on the media. Effective environmental communication requires public engagement in environmental discussions and debates. But if the media is tightly controlled, people might not have access to a range of viewpoints and knowledge that can help them create well-informed beliefs. This lack of candour and openness has the capacity to undermine democracy and impede meaningful dialogue on environmental policy.

Government censorship and media limitations can deter foreign investment and participation in environmental efforts. In today's globalised world, international cooperation is crucial to addressing cross-border environmental concerns. However, by crushing dissent and censoring information, African governments run the danger of turning off potential investors and partners who are hesitant to engage under such restrictive conditions. Obtaining the information and materials required to successfully implement environmental solutions can become more challenging as a result.

5.1.6 Lack of Knowledge

Lack of knowledge hinders people's understanding of the importance of sustainable practices and environmental preservation [8]. In many parts of Africa, people do not have access to formal education or cannot get excellent training that places an emphasis on environmental stewardship and understanding. If people do not have a fundamental understanding of environmental issues, they cannot realise how important it is to preserve the environment for future generations or how their actions influence the natural world. This ignorance can make it more difficult to promote sustainable practices and influence people to adopt more environmentally friendly lives.

Ignorance about environmental risks and concerns can make it more difficult to communicate them effectively. It might be challenging for someone without a basic education to comprehend complex scientific information about environmental issues. This can lead to misconceptions or ambiguity about the causes and impacts of environmental problems, which makes it difficult to motivate the public or decision-makers to take action. It can be challenging for individuals to have a common understanding of environmental issues and their consequences, which is essential for productive communication, in places where there are few educational opportunities.

Myths and misconceptions about the environment can lead to ineffective or harmful responses to environmental challenges, and a lack of knowledge can make them worse. Sometimes ignorance results in actions or beliefs that are detrimental to the environment, such as pollution, habitat destruction, and unsustainable resource extraction. People are at danger of unknowingly contributing to environmental degradation or opposing efforts to implement sustainable solutions if they lack access to trustworthy information and the critical thinking skills necessary to evaluate environmental messages. Education is crucial to dispelling myths and advancing evidence-based tactics that support environmental preservation and resilience.

A lack of education is impeding community involvement and participation in environmental decision-making processes. To effectively address complex environmental challenges, communities must be included in the planning, implementation, and oversight of conservation efforts. On the other hand, those without a basic education cannot be able to effectively contribute to these processes. This can lead to the exclusion of underrepresented groups from environmental decision-making, which could worsen inequality and prevent the development of comprehensive and long-lasting solutions. Education is crucial for empowering people to advocate for their rights to a healthy environment and to actively engage in environmental governance.

5.1.7 Cultural Norms and Values

Many African communities have deeply rooted traditional beliefs about the natural world and the environment [9]. These perspectives often emphasise the spiritual and religious connections that humans have with the natural world. For example, certain African communities consider specific types of vegetation to be sacred and should not be handled or harmed. These opinions can make it difficult for environmental communicators to promote conservation projects that ask for changing or adjusting customary habits.

In many African societies, there are taboos and restrictions on the use of natural resources. For example, it can be illegal to kill or eat certain animals, and it might be illegal to build on or cultivate certain areas of land. These taboos and restrictions can make it more difficult to communicate conservation and sustainable resource management techniques.

In many African cultures, oral history, storytelling, and community gatherings are a few instances of ancient communication methods that are essential to the construction of beliefs and the transfer of knowledge. Conventional communication approaches may not necessarily be consistent with modern environmental communication tactics, which often rely on written materials, digital technology, and scientific data. This division can make it difficult for environmental communicators to effectively connect and engage with local communities.

African cultural practices and beliefs often prioritise immediate survival and needs over the long-term preservation of the environment. For example, in many rural locations, the demands for food, water, and shelter may take precedence over conservation measures. It can become challenging for environmental communicators to promote sustainable behaviours that can require short-term sacrifices in exchange for long-term environmental benefits as a result of this conflict between traditional values and modern conservation aims.

5.1.8 *Denialism About Climate Change*

As part of the denial of climate change, misinformation and incorrect information are often spread [10]. This misleading information has the ability to deceive people and obstruct efforts to raise awareness of the urgent need to address climate change. By sowing uncertainty and ambiguity, climate change sceptics hinder environmental communicators' capacity to clearly convey the seriousness of the issue and mobilise support for long-term solutions.

Denying climate change is undermining conservation efforts in Africa. Denying climate change's reality is deterring individuals and groups from taking action to protect the environment and endangered animals. African ecosystems are suffering greatly from this, as they are already under a lot of stress due to habitat loss, poaching, and climate change.

Besides, Africa is forced to postpone vitally needed mitigation and adaptation efforts because of climate change denial. Some policymakers deny the fact of climate change and are less inclined to fund projects aimed at reducing greenhouse gas emissions, building sustainable infrastructure, and boosting resistance to its consequences. This delay has a severe negative impact on African communities, especially those who are already vulnerable to climate change-related calamities including droughts, floods, and food shortages.

In addition, denying climate change has the potential to undermine efforts to promote sustainable practices and reduce trust in scientific conclusions, so undermining the authority of local people in Africa. Individuals and groups who deny climate change can also deny the experience and direct knowledge of those who are now facing changed weather patterns and declining environmental quality. This is leading to apathy and impotence among the local populace, making it more difficult for them to take action to protect their natural resources and means of survival.

5.1.9 *Lack of Collaboration*

The lack of cooperation among African nations leads to a fragmented approach to environmental issues [11]. Since every country has different priorities and interests, competing policies and strategies exist. This division makes it more difficult to address global environmental issues. Without a concentrated effort, it is practically impossible to make any meaningful progress towards sustainable development and environmental protection.

Moreover, the incapacity to cooperate keeps African countries from sharing knowledge and best practices. Many African countries face similar environmental challenges, so sharing information and experiences can be very helpful. When countries work together, they can learn from each other's successes and failures, which can lead to the creation of environmental policies and programmes that are more effective. However, without collaboration, these opportunities for development and mutual learning are lost, which ultimately impedes the progress of environmental communication and action.

Besides, the absence of collaboration among African nations diminishes their collective impact on the continental arena. When African countries demonstrate unity on environmental issues, they stand a better chance of being taken seriously by the international community. By banding together to promote their interests and priorities, African nations can have an impact on international decision-making processes and win support for environmental initiatives. However, when countries act independently, larger and more powerful nations frequently overshadow them, which diminishes their influence and decreases their ability to effect change.

The lack of cooperation among African nations also exacerbates the differences in environmental communication and action. Smaller and less-developed countries find it challenging to keep up with more developed and resource-rich countries when it comes to environmental issues. If collaborative efforts are not made to close these gaps, disparities in environmental outcomes will persist, impeding the development of a more sustainable future for all Africans.

5.1.10 Lack of Motivation

Poverty and financial difficulties usually take precedence over environmental concerns for a large number of Africans [12]. When resources and opportunities are scarce, individuals prioritise meeting their basic needs above efforts to conserve the environment. This is leading to a lack of motivation and desire to actively engage in initiatives to raise environmental awareness. Since the underlying economic issues that many of them face are not addressed, it is challenging to inspire people to care about the environment and take action.

Africa's low levels of engagement in environmental issues can be attributed to a lack of awareness and information about these issues. It is probable that a large number of populations are ignorant of the importance of conservation efforts or how their actions influence the environment. If people do not have access to trustworthy information and resources, they cannot see the value of environmental communication or be motivated to participate.

In Africa, people's interactions with environmental communication can also be influenced by cultural values and beliefs. Some communities follow traditions or ideologies that are in opposition to the state of conservation efforts. This is obstructing efforts to promote sustainable behaviour and make communication more challenging. In addition, a lack of trust in institutions and governmental agencies is another reason for Africa's low participation rates in environmental communication. Many people mistrust the reasons for environmental initiatives, or they do not believe that their actions can have an influence.

5.1.11 *Cultural Differences and Beliefs That Influence Environmental Communication*

Traditional African beliefs often emphasise the interconnectedness of humans and the natural world [13–16]. More reverence for the environment arises from this, but it also gives rise to the notion that environmental deterioration is not human-caused. It is difficult to convey the urgency of tackling environmental issues since some communities do not believe that these are their problems to solve.

In addition, the variety of languages and dialects spoken across the African continent are a significant barrier to effective environmental communication. Different communities hold different cultural perspectives on environmental issues, which make it challenging to convey a message that is understood by everyone. Because of this linguistic diversity, there can be misunderstandings and poor communication, which makes it more challenging to raise public awareness and mobilise support for pressing environmental issues.

Moreover, there are times when recognised environmental conservation strategies conflict with African cultural traditions and customs. For example, some communities rely on hunting, fishing, and farming practices that could be detrimental to the environment. Attempting to explain the need for these practices to change can be met with resistance and scepticism because they are deeply rooted in cultural identity and traditions. Initiatives for sustainable practices and conservation can be hampered by this opposition, worsening the state of the environment.

Additionally, the Western and colonial ideologies that influenced African cultures also have an effect on environmental communication. Effective engagement and participation is hampered by the beliefs and values of African communities not always aligning with Western narratives and solutions to environmental problems. Ignoring local cultural contexts and imposing Western approaches of sustainability and conservation can undermine the effectiveness of environmental communication campaigns.

5.1.12 *Misinformation and Disinformation*

Misinformation and disinformation undermine conservation efforts by spreading false or misleading information about environmental issues. False claims about the effects of climate change or deforestation, for example, lead to misguided policies and actions that exacerbate environmental problems rather than solve them. Ultimately, this has detrimental impact on the ecosystems and natural resources that are essential to the survival of many African communities [17].

As a result of false information and deceit about environmental issues, people and communities in Africa are gravely at risk for significant health difficulties. For example, disseminating false information on the safety of particular chemicals or pesticides can expose individuals to these substances and result in long-term health problems. False information about the quality of the air or water in a certain area also deters individuals from taking the necessary precautions to protect themselves from harm.

Inaccurate environmental information has detrimental economic effects on African communities. For example, inaccurate information about the sustainability of specific industries or activities has an impact on poor decision-making that puts those whose livelihoods depend on natural resources in financial difficulties. Furthermore, opportunities for sustainable economic development and progress are missed as a result of misleading information about the consequences of climate change or other environmental issues.

Misinformation and disinformation have the potential to widen social divides in African communities, particularly in relation to environmental issues. False narratives about the causes of environmental deterioration or the allocation of resources, for example, can fuel tension and conflict among different groups in a society. It can be difficult to address environmental issues and find long-term solutions that benefit the community as a whole as a result.

5.1.13 Lack of Governmental Commitment and Will

Projects aimed at promoting environmental awareness lack the funding and support necessary to be successful in the lack of governmental commitment and will [18]. Government officials are essential in the process of creating environmental laws and regulations that can help solve pressing environmental issues. When environmental groups and activists do not prioritise environmental communication, it becomes more difficult for them to access funding, information, and other essential resources. This lack of support slows down the settlement of environmental concerns and hinders the effectiveness of environmental communication activities.

Politicians that do not think environmental communication is important cannot see the advantages of informing the public about environmental issues. As a result of lack of political will and dedication behind campaigns, workshops, and other initiatives that educate the public about the need of environmental protection, government officials have little incentive to support them. This means that a sizeable portion of Africans continue to be unaware of the impact of their actions on the environment and fail to take appropriate steps to lessen their environmental footprint.

A lack of political will and dedication to environmental communication encourages apathy and indifference towards environmental problems. When government leaders do not actively participate in environmental communication initiatives, it appears as though the environment is not a major issue. This lack of leadership leads to inaction on environmental issues and public scepticism, which makes it much more difficult to win support for conservation and sustainable development initiatives.

Without political will and commitment, environmental communication efforts do not have the necessary accountability measures in place to ensure that environmental problems are being handled in a way that promotes society. As a result, environmental degradation goes unchecked and endangers ecosystems and natural resources irreversibly because environmental communication is not prioritised.

5.2 Opportunities for Environmental Communication in Addressing Environmental Issues

Africa has a plethora of significant opportunities for environmental communication to address environmental issues. Stakeholders can be empowered, informed, educated, and mobilised via good communication, allowing people on the continent to work together to protect the environment.

- **Public awareness campaigns:** Environmental communication can help to address environmental issues by educating the public about environmental issues and inspiring them to take action to protect the environment. These campaigns can be implemented via a range of media, such as radio, social media, television, and community outreach programmes [19].
- **Community engagement:** Environmental communication can be employed to influence vigorous community involvement in promoting sustainable environmental practices. Encouraging communities to take ownership of environmental projects and participate in decision-making processes can foster a sense of accountability and dedication to environmental protection [20–22].
- **Environmental education programmes:** Establishing programmes for environmental communication education in schools can help cultivate a sense of environmental stewardship in the next generation. When learners understand the importance of conservation and sustainable practices, they can grow up to be environmental advocates.
- **Collaborative partnerships:** Environmental communication can help to influence a variety of stakeholders to work together, such as companies, non-governmental organisations, and governmental agencies that can lead to the development of comprehensive environmental communication strategies. Cooperation to accomplish common goals can have a greater positive impact on the environment.
- **Promotion of eco-tourism:** As a result of its abundant natural environments and varied wildlife, Africa is a top destination for eco-tourism. The positive environmental impacts of tourism can be amplified through the environmental communication endorsement of environmentally conscious travel tactics and conscientious traveller conduct [23].
- **Corporate responsibility:** Environmental communication can foster corporate responsibility by promoting moral business conduct and making companies answerable for their environmental impact. By highlighting

sustainability and promoting transparency, communication can promote environmental responsibility and corporate responsibility.

- **Behaviour change:** Communication campaigns that also encourage sustainable, ecologically friendly behaviour can centre on individual behaviour. Environmental communication can be used to support programmes like recycling, waste reduction, and energy conservation by influencing behaviour change at the individual and community levels.
- **Technology and innovation:** Through environmental communication, cutting-edge innovations and technologies that support sustainable development in Africa can be emphasised. By presenting successful strategies and case studies, communication can inspire people to come up with original solutions to environmental issues.
- **Monitoring and evaluation:** Environmental communication can help keep tabs on and monitor and evaluate how well environmental initiatives are working in terms of their effects on the environment. Organisations can gather information and feedback via communication channels to assess the success of their programmes and decide on the best course of action.
- **Youth engagement:** Through environmental communication, young people in Africa can be empowered to lead as the next generation of environmental leaders. By providing a forum for young voices and elevating their perspectives, communication can support youth-led projects and effect meaningful change [24].
- **Policy influence:** Environmental communication has the power to influence the African environmental agenda. Engaging policymakers, the media, and the public through effective communication can advance sustainable policies that prioritise environmental conservation and protection.
- **Promoting sustainable behaviour:** Through targeted messaging and campaigns, environmental communication can promote sustainable behaviours like energy conservation, recycling, and responsible consumption. By encouraging people to make little changes in their daily lives, environmental communication can all together have a big impact on the environment.
- **Empowering local communities:** Communities can be equipped with the knowledge and skills necessary to take responsibility for their natural resources and the environment by means of environmental communication. Resilience and sustainability can be built from the ground up by providing communities with the knowledge and abilities they need to protect their land and water.
- **Promoting green technologies:** Through communication channels like the media, social media, and public events, environmental communication can raise public awareness of sustainable practices and green technologies that can help to lessen pollution and climate change. By disseminating success stories and best practices, environmental communication can motivate individuals to utilise these technologies in their daily lives.

- **Encouraging a conservation culture:** Environmental communication can help African societies develop a conservation culture and a respect for the natural world. By acknowledging the beauty and importance of the continent's natural heritage, environmental communication can foster a sense of pride and responsibility in preserving it for future generations.
- **Addressing environmental justice:** Environmental communication can draw attention to the ways that environmental degradation disproportionately affects marginalised communities in Africa. By giving voice to those most affected by environmental injustices, environmental communication can advocate for policies and laws that promote environmental equity and social justice.
- **Sparking collective action:** Effective environmental communication ultimately has the ability to motivate individuals, communities, and groups to come together and take on environmental challenges in Africa. By fostering a sense of collective urgency and responsibility, environmental communication can work to create a more resilient and sustainable future for the continent [25].
- **Mobile and digital penetration:** The first major opportunity lies in leveraging the continent's rapidly expanding mobile and digital penetration. Mobile-first environmental campaigns can disseminate real-time alerts about deforestation, pollution events, or impending weather hazards, significantly enhancing early warning systems [19].
- **Integrating Indigenous and local knowledge:** There is a significant opportunity in integrating Indigenous and local knowledge into modern environmental narratives. Traditional ecological wisdom, often preserved through oral traditions in communities such as the Maasai or the San people, offers context-specific, time-tested solutions for resilience that resonate deeply with local populations, creating more culturally relevant communication strategies [20].
- **Professionalisation of environmental journalism:** Many African nations lack specialised science reporters, leading to sensationalism or outright ignorance regarding complex issues like biodiversity loss or renewable energy transitions. Investing in training journalists to accurately report on climate science and policy negotiations creates an informed public discourse, moving beyond simplistic narratives [21].
- **Utilising radio:** Radio, which remains the most accessible medium in many rural areas, presents a powerful opportunity for community-based dialogue. Interactive radio programmes can facilitate two-way communication, allowing citizens to voice concerns about local pollution or resource depletion directly to decision makers [19].
- **Case studies:** There is an opportunity in crafting powerful narratives around successful adaptation and mitigation case studies. Highlighting localised successes, such as community-managed solar microgrids in Ghana or successful mangrove restoration projects in Nigeria, provides tangible evidence that change is possible, thereby combating widespread climate fatalism [22].

- **Environmental communication training:** Establishing dedicated environmental communication training centres within African universities is another opportunity. These centres can foster a new generation of practitioners skilled not only in scientific communication but also in cross-cultural mediation and crisis communication specifically tailored to the African context [19].

5.3 Conclusion

Environmental communication in Africa faces numerous challenges such as limited access to information, language barriers, cultural differences, and lack of awareness on environmental issues. Political instability and corruption also hinder effective communication among stakeholders. However, there are opportunities for progress and success. The growing recognition of environmental sustainability among governments, businesses, and civil society has led to the development of policies and initiatives for sustainable development. Advances in technology, such as mobile phones and the Internet, have made it easier to disseminate information and engage the public in environmental issues. Social media platforms have also provided a platform for individuals and organisations to raise awareness and mobilise support for environmental causes. By leveraging these opportunities, Africa can work towards a more sustainable future.

Recap

While environmental communication in Africa faces significant obstacles due to linguistic diversity, socioeconomic challenges, and competing priorities, it also presents remarkable opportunities for positive change. By adopting culturally sensitive approaches, leveraging technology, and fostering collaboration, it is possible to enhance understanding of sustainability and environmental protection across the continent. Through decisive action and inclusive dialogue, Africa can pave the way for a more sustainable future.

Questions

- What are the messages of environmental sustainability that need to be communicated in Africa?
- What are the messages of environmental protection that need to be communicated in Africa?
- What are the challenges to environmental communication in Africa?
- What are the opportunities to environmental communication in Africa?

References

1. Types of environmental issues: meaning, pollution, videos, examples. Toppr-guides. 2018.
2. Obi CI. Environmental movements in sub-Saharan Africa: a political ecology of power and conflict. UN Research Institute for Social Development; 2005.
3. Bosetti V, Lubowski R, editors. Deforestation and climate change. Edward Elgar; 2010.
4. Banda Jr, S. Deforestation reaches worrying level—UN. africanews.com. 2018. https://web.archive.org/web/20081008083104/http:/www.africanews.com/site/list_messages/18831.
5. Roe S, Streck C, Beach R, et al. Land-based measures to mitigate climate change: potential and feasibility by country. Glob Chang Biol. 2021;27(23):6025–58.
6. Deforestation statistics for Nigeria. Mongabay. 2021. https://worldrainforests.com/deforestation/archive/Nigeria.htm.
7. Wana F, Senapathy M. Small-scale irrigation utilisation by farmers in Southern Ethiopia: monograph. Primedia eLaunch LLC; 2023. p. 174. https://doi.org/10.36074/ssiubfise-monograph.2022.
8. Oyediji OT, Adenika OA. Forest degradation and deforestation in Nigeria; poverty link. Int J Multidiscip Res Anal. 2022;5(10):2837–44.
9. Hillstrom K, Hillstrom LC. North America: a continental overview of environmental issues. USA: Bloomsbury Publishing; 2003. p. 85–6.
10. Favretto N, Dougill AJ, Stringer LC, Afionis S, Quinn CH. Links between climate change mitigation, adaptation and development in land policy and ecosystem restoration projects: lessons from South Africa. Sustainability. 2018;10(3):779.
11. IFA: International Fertilizer Industry Association. Soil degradation in Africa Low use of fertilizers is one of the main causes of environmental degradation. Fertilizer.org. IFA; 2013. https://web.archive.org/web/20130510070935/http:/www.fertilizer.org/ifa/HomePage/SUSTAINABILITY/Soil-degradation-in-Africa.
12. Land and environmental degradation and desertification in Africa. FAO; 2013. https://www.fao.org/4/X5318E/x5318e02.htm.
13. IFA: International Fertilizer Industry Association. Soil degradation in Africa low use of fertilizers is one of the main causes of environmental degradation. Fertilizer.org. IFA; 2013.
14. Soil degradation. Goodplanet.info; 2009. https://web.archive.org/web/20130518084447/http:/www.goodplanet.info/eng/Pollution/Soils/Soil-degradation.
15. Lanly JP. Tropical forest resources food and agriculture organization of the United Nations. Rome, Italy: United Nations; 1982. http://www.fao.org/docrep/015/an778e/an778e00.pdf
16. Wood TS, Baldwin S. Fuelwood and charcoal use in developing countries. Annu Rev Energy. 1985;10(1):407–29.
17. Jacobson MZ. On the causal link between carbon dioxide and air pollution mortality. Geophys Res Lett. 2008;35(3):S2CID 16440166. https://doi.org/10.1029/2007GL031101.
18. World Health Organisation. Regional burden of disease due to indoor air pollution; 2013. https://web.archive.org/web/20040701221414/http:/www.who.int/indoorair/health_impacts/burden_regional/en/.

19. Cukrowska EM, Yalala BN, Tutu H, Chimuka L. Spatial distribution and source identification of mercury in dust affected by gold mining in Johannesburg, South Africa. In: Proceedings of the international conference of recent trends in environmental science and engineering; 2017. p. 104.
20. Lusilao-Makiese JG, Cukrowska EM, Tessier E, Amouroux D, Weiersbye I. The impact of post gold mining on mercury pollution in the West Rand region, Gauteng, South Africa. J Geochem Explor. 2013;134:111–9.
21. Papu-Zamxaka V, Mathee A, Harpham T, et al. Elevated mercury exposure in communities living alongside the Inanda Dam, South Africa. J Environ Monit. 2010;12(2):472–7.
22. Liousse C, Assamoi E, Criqui P, Granier C, Rosset R. Explosive growth in African combustion emissions from 2005 to 2030. Environ Res Lett. 2014;9(3):035003.
23. Guide on ambient air quality legislation – air pollution series. United Nations Environment Programme; 2023. https://doi.org/10.59117/20.500.11822/42536. ISBN 978-92-807-4029-5.
24. Scorgie Y. Air quality and regulation. National Association for Clean Air; 2013.
25. Kang H, Zhu B, Zhu C, de Leeuw G, Hou X, Gao J. Natural and anthropogenic contributions to long-term variations of SO2, NO2, CO, and AOD over East China. Atmos Res. 2019;215:284–93. https://doi.org/10.1016/j.atmosres.2018.09.012. ISSN 0169-8095

CHAPTER 6

Stakeholder Engagement in Environmental Communication

Introduction
Stakeholder management is a need for any organisation's operations, but it is crucial for environmental communication in Africa in particular. Finding, assessing, and communicating with individuals or groups involved in environmental communication are all necessary. Stakeholders must be successfully engaged in environmental communication for there to be credibility, support, and ultimately long-term success. Given stakeholders' significant effect on public opinion and decision-making around environmental issues, it is important to comprehend the diverse range of stakeholders involved in environmental communication. This chapter's major focus is on stakeholders' engagement in environmental communication. The chapter begins with a description of stakeholder's engagement and examines its importance in environmental communication. The chapter then discusses the many categories of stakeholders in environmental communication. The chapter ends by analysing current disputes among stakeholders and potential solutions to these conflicts.

6.1 The Concept of Stakeholder Engagement

The term "stakeholder engagement" is traced back to the mid-twentieth century when management scholars began to recognise the importance of stakeholders in influencing organisational decisions [1]. The term "stakeholder" is a combination of the words "stake" and "holder." An individual who has a direct interest, or "stake," in the outcomes and performance of a particular organisation is referred to as a stakeholder [2–4]. The term "stakeholder" now encompasses a broader range of individuals and groups that either are affected

© The Author(s), under exclusive license to Springer Nature Switzerland AG 2025

G. Mutinta, *An African Perspective on Environmental Communication*,
https://doi.org/10.1007/978-3-032-05052-6_6

by or affect the choices and actions of an organisation. This includes not only shareholders and investors but also employees, customers, suppliers, the neighbourhood, and even the environment.

The term “engagement” is often used to describe a period of time when a person is actively involved in a cause or group, or when a couple has made the decision to be married. The word “engagement” comes from the Old French word “engagier,” which means to promise or commit [5]. This word comes from the Latin “individuus,” which means “indivisible.” In essence, a person who is engaged to marry is making a life-long commitment to their partner, promising them faithfulness and devotion.

However, engagement is not limited to romantic relationships. It could also imply a commitment to or involvement with a cause, project, or organisation. When someone is engaged in their work, they are making a positive impact on the world, actively contributing, and exerting great effort to achieve their goals. This sort of commitment is crucial in the present Africa, when there are a plethora of social, political, and environmental issues competing for people’s attention.

Engaging in causes one is passionate about enhances society and provides people’s lives purpose and satisfaction. People’s happiness and personal growth depend on them doing important things, like volunteering at a local charity, contributing to organisations people support, or simply showing up and paying attention when people are with those they love.

Stakeholder engagement, therefore, involves the process of building relationships with key stakeholders to understand their perspectives, needs, and concerns, and to collaborate with them in decision-making processes. This approach has become increasingly essential in Africa’s complex and interconnected organisational environment, where organisations are under pressure to consider a wide range of interests beyond just shareholders.

In the realm of environmental communication, understanding the different types of stakeholders and their levels of power, urgency, and legitimacy is crucial in order to effectively engage with them. The concept of power, urgency, and legitimacy is often used to assess the importance and influence of stakeholders [6–10]. Power refers to the ability of stakeholders to influence decisions and outcomes, while urgency relates to the time constraints and importance of a stakeholder’s concerns. Legitimacy, on the other hand, refers to the perceived validity and credibility of a stakeholder’s claims and interests. By understanding these elements, organisations can prioritise their engagement with stakeholders and tailor their communication to meet their needs.

There are three main types of stakeholders that organisations must consider in environmental communication: latent stakeholders, expectant stakeholders, and definitive stakeholders [11–15]. Latent stakeholders are those who are not yet engaged or aware of their interests in a particular issue, but can become involved if their concerns are brought to light. Expectant stakeholders are those who have a vested interest in the issue but may not have the power or urgency to influence decisions. Definitive stakeholders, on the other hand, are those who have power, legitimacy, and urgency to impact decisions and outcomes.

When engaging with stakeholders in environmental communication, it is important to consider their levels of power, urgency, and legitimacy in order to tailor the messaging and engagement strategies accordingly. By identifying and prioritising stakeholders based on these factors, organisations can build stronger relationships, address concerns more effectively, and ultimately drive positive change for the environment.

Since environmental issues are becoming increasingly widespread and serious in Africa, it is progressively evident that in order to solve environmental challenges, action must be taken at all levels. These concerns will only intensify. This is where stakeholder engagement in environmental communication programmes comes into play.

6.2 The Importance of Stakeholder Engagement in Environmental Communication

Including stakeholders can help spark innovative solutions to environmental issues by bringing together a variety of perspectives, backgrounds, and specialties [16–21]. Including stakeholders in environmental communication can increase the credibility and reliability of the messages. When stakeholders are actively involved in the decision-making process and have a part in creating environmental policies and strategies, they are more likely to accept the proposed measures and believe the facts being provided.

Stakeholder engagement can play a critical role in improving environmental decision-making processes. Policymakers can create decisions that are more likely to be approved and successfully carried out by consulting with key stakeholders to better understand their preferences, concerns, and interests. Involving stakeholders in environmental communication can promote more accountability and transparency in the decision-making process. When addressing environmental challenges, decision-makers can demonstrate their commitment to transparency, adaptability, and inclusivity by including stakeholders in talks, meetings, and deliberations.

Besides, stakeholder engagement promotes cooperation and collaboration despite the fact that multiple stakeholders can have diverse objectives and areas of interest [22–26]. By bringing together a range of stakeholders and encouraging conversation and information sharing, environmental communication can promote improved understanding, consensus-building, and cooperative action towards shared environmental goals. Public awareness of environmental issues can rise and people's meaningful engagement in environmental communication can be achieved by including stakeholders. By including stakeholders in outreach initiatives, educational initiatives, and community events, environmental communication can empower individuals to take action, make informed decisions, and contribute to positive environmental outcomes.

In addition, stakeholder engagement is essential to ensuring the resilience and long-term sustainability of environmental communication projects. By building strong networks, relationships, and partnerships with stakeholders, environmental communication can eventually establish the foundation for long-term collaboration, adaptation, and development in resolving environmental challenges.

6.3 Stakeholders

Environmental communication is essential for raising awareness about the importance of preserving our planet and achieving sustainable living practices. It is essential to identify and include important stakeholders in the communication process in order to reach a larger audience. Stakeholders are people or organisations with a stake in or influence over environmental concerns [27–30]. They are important because of their capacity to influence public opinion and decision-making.

6.3.1 Non-governmental Organisations

Environmental issues are posing a serious risk to Africa and its future generations [31]. Non-governmental organisations are therefore crucial for encouraging environmentally friendly behaviour and sustainable methods. Through environmental communication, non-governmental organisations can draw the public's attention to pressing environmental issues. Organisations can use a range of platforms, such as social media, public events, and educational programmes, to disseminate information and educate the public on the significance of environmental problems. Raising public awareness is essential to generating interest in and encouraging action towards positive environmental change.

Environmental communication helps create a sense of community and shared responsibility for environmental concerns. Through public interaction and conversation, non-governmental organisations can foster collaboration and a sense of shared purpose in the endeavour to address environmental concerns. The public's involvement in environmental activities can be stimulated and non-governmental organisations' impact can be increased by this sense of belonging.

Through environmental communication, non-governmental organisations can influence policy changes and decision-making at many levels. Non-governmental organisations can effectively promote the implementation of sustainable policies and practices by showcasing the opinions of experts, supporting data from science, and real-world examples of the effects on the environment. This advocacy work is crucial to achieving environmental justice and structural change for underprivileged communities.

Besides, environmental communication facilitates the mobilisation of resources and support for non-governmental organisations programmes and operations. By effectively communicating their vision, goals, and effects, environmental organisations can attract donors, volunteers, and other stakeholders. This support is essential for non-governmental organisations to effectively carry out their work and have a substantial influence on the fight against environmental degradation.

6.3.2 *Governmental Agencies*

In an increasingly urgent continent of Africa where environmental concerns are gaining priority, government agencies play a critical role in educating the public and inspiring people to take part in efforts to preserve and protect our planet [32].

Environmental communication facilitates public education and awareness-building around significant environmental challenges [32]. Governmental agencies can help individuals make better decisions about their everyday activities and lifestyle choices by disseminating knowledge about critical environmental issues such as climate change, biodiversity loss, and air and water pollution. This increased understanding can lead to widespread support for policies and initiatives that promote sustainability and protect the environment for coming generations.

Adequate environmental communication can foster accountability and transparency within government agencies. By being open about their environmental policies, practices, and goals, government agencies can build confidence with the public and demonstrate their commitment to environmental stewardship. Transparency can assist to ensure that environmental concerns continue to be given top priority when drafting public policy by enticing people to hold public authorities accountable for their environmental decisions and actions.

Additionally, effective environmental communication can facilitate meaningful dialogue and cooperation between government agencies, the general public, and stakeholders. Governmental organisations can settle conflicts, get feedback, and incorporate a variety of perspectives into their environmental planning and decision-making processes by using interactive communication platforms. This inclusive approach to consensus-building and conflict resolution can help produce longer-term, more equitable, and sustainable solutions.

As a result of environmental communication, public support and action for sustainability and environmental protection initiatives can grow. By addressing environmental issues in ways that appeal to people's values, aspirations, and beliefs, governmental agencies can inspire citizens in Africa to take meaningful steps to reduce their impact on the environment, support conservation efforts, and demand improved environmental protections.

6.3.3 *Civil Society Organisations*

Civil society organisations can play a crucial role in advocating for environmental issues and promoting sustainable behaviours in our society [33–35].

Effective environmental communication is essential to bringing serious environmental problems to the public's attention. Civil society organisations can be in the forefront of examining and documenting environmental risks. These organisations can use a range of media, such as social media, public campaigns, and educational programmes, to properly highlight these issues and inform the public about the urgent need to act immediately to protect our world.

Through the use of compelling stories, data, and imagery, civil society organisations can inspire individuals to take action, such as signing a petition, participating in a demonstration, or embracing more sustainable lifestyles. By actively engaging with the public, these organisations can cultivate a loyal following of individuals who are dedicated to promoting environmental sustainability and conservation.

Civil society organisations can play a major role in supporting corporate responsibility initiatives, stronger environmental regulations, and sustainable business practices in a variety of industries. By effectively conveying the environmental consequences of certain policies and practices, these organisations can sway politicians to take action and implement changes that benefit the environment and people.

In addition, Civil Society Organisations can highlight the interconnectedness of all living things and the need of protecting natural resources. This can help to promote an understanding of environmental responsibility. Through engagement, storytelling, and education, civil society organisations can empower individuals to make informed decisions that benefit the environment and the next generation.

6.3.4 *Researchers*

Environmental comprehension is facilitated by public-scientist communication [36]. By successfully publicising their findings, researchers can enlighten the public about important environmental concerns; researchers might inspire people to take action. Increased involvement, support, and awareness of environmental concerns can come from this.

Building credibility and trust within the scientific community can be facilitated via environmental communication. In relation to environmental issues, misinformation and scepticism about science have increased recently. By openly disclosing their research procedures, conclusions, and ramifications, scientists can build public confidence and advance an open research environment. This can lead to increased acceptance of scientific consensus and environmental protection laws.

Environmental communication can influence policy and decision-making. Research results have a major role in shaping environmental legislation and policies. By effectively communicating their findings to stakeholders and policymakers, researchers can influence the development of scientifically grounded environmental policies and encourage evidence-based decision-making. Better long-term and effective solutions to environmental issues could arise from this.

In addition, environmental communication can inspire action and behavioural changes. Research findings can encourage individuals, communities, and organisations to adopt environmentally friendly practices and support conservation efforts. By effectively communicating the importance of environmental issues and the role that individuals can play in reducing them, educators can inspire action and promote positive changes. This can motivate teamwork to protect and preserve the ecology.

6.3.5 *Community Leaders*

Africa is plagued by several issues, and effective solutions need collaboration amongst all segments of society. Community leaders are crucial because they can inspire action and increase awareness in their communities [37]. For example, local leaders are well positioned to engage with a diverse range of individuals within their community. Among other stakeholders, they can communicate with corporations, people, and public personnel. By taking part in environmental communication, community leaders can educate and inspire others to take action for environmental conservation. They can raise awareness of local environmental issues, encourage community members to get involved in conservation efforts, and propagate sustainable practices.

Community leaders can serve as role models for their fellow residents by demonstrating their commitment to environmental conservation. When community leaders actively engage in environmental communication and activities, they can set an example for others to follow. Leading by example, community leaders can inspire others to adopt sustainable habits. These behaviours include by reducing litter, conserving electricity and water, and funding environmentally friendly initiatives. Their actions can have a domino effect, encouraging the neighbourhood to adopt an ecologically responsible mindset.

Moreover, community leaders have the power to influence policies that might significantly impact the environment. By being active in environmental communication, community leaders can support laws and policies that protect natural resources, preserve wildlife habitats, and slow down climate change. They can work together with community members, organisations, and local government to address environmental injustices, promote sustainable development, and improve the quality of the environment. Through effective advocacy and communication, community leaders can foster positive change and build a more sustainable future for their communities.

Community leaders have the capacity to foster collaboration and partnerships that fortify efforts aimed at protecting the environment. Engaging in environmental communication can help community leaders build relationships with other stakeholders, such as companies, non-profits, schools, and community organisations. To collaboratively address environmental problems, they can organise activities, share resources, and collaborate on initiatives. By uniting, community leaders can increase their impact, take use of their advantages, and enhance the environment.

6.3.6 Faith-Based Organisations

Faith-based organisations in Africa hold a unique position in society due to their ability to inspire and encourage people to make positive changes in the world [38]. Environmental communication is a key area in which they can have an effect.

Faith-based organisations have an ethical obligation to preserve the environment. Respect for and preservation of the environment are fundamental to many religious doctrines. By engaging in environmental communication, faith-based organisations can serve as a reminder to its members of their responsibility to protect the environment for coming generations. Faith-based organisations can be useful in projects like recycling campaigns, environmental protection efforts, and encouragement of sustainable living.

Africa has a large presence of faith-based organisations that have a big effect on local communities. They can motivate a substantial number of people to strive for a common goal. By engaging in environmental communication, faith-based organisations can motivate individuals to take action and draw attention to pressing environmental issues. This can lead to changes in behaviour, legislation, and lifestyle choices that are advantageous for the environment.

Faith-based organisations can offer a unique perspective to the environmental debate. The importance of peacefully coexisting with nature and the interdependence of all living things are emphasised by many religious traditions. Through their involvement in environmental communication, faith-based organisations can help spread these teachings and provide a spiritual basis for understanding and addressing environmental concerns. This has the power to motivate individuals to make major efforts to protect the environment and to view environmental stewardship as a sacred duty.

By engaging in environmental communication, faith-based organisations can successfully advocate for environmental justice and build relationships with other groups. By collaborating with other like-minded groups, faith-based organisations can share their message and work towards common goals. This can encourage a sense of togetherness and mutual support across diverse groups, which can lead to more important and long-lasting solutions to environmental concerns.

6.3.7 Media Organisations

Media organisations are crucial for forming opinions, disseminating information, and influencing public policy [39].

Media sources have a big influence on people in Africa. By utilising a range of media, such as social media, radio, newspapers, and television, they can spread their messages and provide information about environmental concerns to a large audience. By using their platforms to promote environmental preservation and sustainable practices, media outlets can raise public awareness and spur action.

Many people in Africa do not have access to formal environmental education. By providing resources, information, and education on the environmental difficulties the continent is facing, media outlets can be able to narrow this gap. Through documentaries, news broadcasts, and educational initiatives, media outlets can help inform the public about the importance of biodiversity, climate change, pollution, and other pressing environmental issues.

Media outlets can support and monitor environmental conservation efforts in Africa. By highlighting environmental violations, endorsing sustainable practices, and holding governments and corporations accountable for their actions, media outlets can promote positive change and push for more stringent environmental regulations. By featuring success stories and the efforts of local communities and conservation organisations, media sources can also inspire individuals to take action and support environmental conservation programmes.

Besides, media organisations have the power to influence legislation and change public opinion. By covering environmental issues and pushing for stronger environmental laws and regulations, media outlets can help create a more conducive atmosphere for sustainable development in Africa. Media source can actively participate in environmental communication to ensure that environmental activists' perspectives are heard, and valued, as public opinion and media coverage regularly influence politicians.

6.3.8 Individuals at the Grassroots Level

Every community starts with its grassroots members, who play a crucial role in shaping opinions, effecting change, and creating values [40]. For these reasons, it is essential that Africans at the grassroots level are involved in environmental communication.

At the grassroots level, people are acutely aware of the issues that affect their neighbourhood and the surrounding areas. When it comes to environmental issues, they bear the brunt of the consequences. By utilising the expertise and experience of those working at the grassroots level in environmental communication, environmental communication can develop more effective strategies for addressing these issues. Their own experience and observations can help us identify the root causes of environmental problems and create long-term, community-focused solutions.

Individuals at the local level have a significant impact on their communities. They can mobilise grassroots movements, motivate their peers, and raise public awareness of environmental issues. By including grassroots people in environmental communication, environmental communication can use their networks and influence to reach a wider audience. Since grassroots activists are often trusted sources of information in their communities, they are a valuable resource for increasing awareness and encouraging behavioural change.

Besides, promoting community involvement in environmental communication can help bridge the communication gap between local residents and government representatives. By building relationships, encouraging communication, and advocating for ecologically sound policies that are advantageous to all sides, activists on the ground can serve as a link between the government, non-governmental groups, and local citizens. By including grassroots people in environmental communication, campaigns can ensure that the needs and views of marginalised groups are taken into account when decisions are being made that affect their environment.

Participation of grassroots individuals in environmental communication can foster a sense of empowerment and ownership among communities. Engaging in debates about environmental issues at the grassroots level increases the likelihood that people will feel compelled to protect the environment and take action to address environmental issues. By providing common people with the means to care for their surroundings, environmental communication can foster a culture of environmental consciousness and effect long-lasting change that benefits people and the environment.

6.4 Conflicts Among Environmental Advocates

Conflicts among environmentalists are prevalent as a consequence of opposing ideologies, goals, and conservation strategies [41, 42]. These conflicts may hinder progress towards unified aims and address significant environmental challenges.

6.4.1 Industry Versus Environmentalists

Conflicting interests prevail among environmental stakeholders on this critical objective, particularly between industry representatives and environmentalists [42].

Industry stakeholders in Africa, particularly those in the mining, agriculture, and manufacturing sectors, primarily concentrate on profit maximisation and operational growth. They often prioritise economic growth above environmental concerns, asserting that economic expansion is essential for social well-being. This sometimes leads to environmentally harmful acts, such as pollution, deforestation, and resource depletion. In their quest for profit, industrial stakeholders may overlook the long-term impacts of their operations on the environment, leading to harmful consequences for ecosystems and natural resources.

In contrast, environmentalists advocate for environmental preservation and the promotion of sustainable practices [43–47]. They are concerned about the consequences of industrial activities on the environment, such as air and water pollution, habitat deterioration, and climate change. Environmentalists assert that the exploitation of natural resources and the degradation of ecosystems are unsustainable over time, requiring urgent measures to preserve the environment for future generations. They often promote more rigorous regulations and policies to alleviate the environmental impact of industrial operations and foster sustainable practices.

The conflicting interests of industrial stakeholders and environmentalists obstruct the formation of consensus and the implementation of effective environmental rules. Industry stakeholders may resist legislation that restricts their operations, apprehensive about its adverse effects on profitability and competitive advantage. Conversely, environmentalists may see industrial players as adversaries prioritising profit above the well-being of the planet and its inhabitants. This leads to clashes, debates, and unrest from both parties, hindering the achievement of agreement on the best strategy to preserve the environment while promoting economic development.

- **Solution**

To reconcile these conflicting interests, it is essential for industrial stakeholders and environmentalists to engage in dialogue and collaboration to discover mutual benefits and solutions that serve both ecological and economic objectives. Industry stakeholders may use sustainable practices that reduce their environmental effect and contribute to the conservation of natural resources. They may also invest in sustainable technologies and ideas to mitigate pollution and waste generation. Environmentalists can partner with industrial stakeholders to raise awareness about the need of environmental protection and advocate for legislation that promote sustainable development.

6.4.2 *Government Versus Citizens*

The protection of the environment and the promotion of sustainable practices in Africa are critical issues requiring the involvement of many stakeholders, including government entities and the general public. These groups often display conflicting interests about the prioritising of environmental conservation versus economic development [48].

African nations often encounter the challenge of balancing environmental preservation with economic development. Governments must ensure the sustainable use of natural resources and protect the environment for future generations. Conversely, they have pressure to foster economic growth and provide jobs for their population. This leads to conflicts of interest, since governments tend to prioritise short-term economic gains above long-term environmental sustainability.

Governments are predisposed to provide mining or logging permits to businesses to foster economic growth, even at the detriment of vital ecosystems. In some cases, governments disregard illegal logging, poaching, or pollution to attract foreign investment and create jobs. This leads to disastrous consequences for the ecology and local people, resulting in deforestation, loss of biodiversity, and environmental degradation.

Conversely, citizens in Africa often endure the consequences of environmental degradation and unsustainable practices. Local communities rely on natural resources for their subsistence, including agriculture, fishing, and hunting. The exhaustion or contamination of these resources may profoundly affect their well-being and economic stability.

Citizens often suffer the repercussions of environmental degradation, although perhaps lacking representation in decision-making processes. In several African countries, citizens have limited access to information and resources essential for championing their environmental rights. This impedes their ability to hold governments and corporations accountable for their actions, leading to less transparency and responsibility in environmental governance.

- **Solution**

Improved collaboration and cooperation between governments and people in Africa is crucial to resolve conflicting interests on environmental protection and sustainable practices. Governments must prioritise the long-term sustainability of the environment and the well-being of their populations above immediate economic gains. This requires stringent environmental regulations, effective enforcement mechanisms, and public participation in decision-making processes.

Citizens must be empowered to advocate for their environmental rights and hold governments and corporations accountable for their actions. This may be achieved via education, awareness improvement, capacity-building, and access to information and resources. By collaborating, governments and people can promote sustainable practices, protect the environment, and ensure a greener and more prosperous future for everybody in Africa.

6.4.3 Local Communities Versus Conservationists

In Africa, environmental stakeholders have divergent interests about environmental conservation and the advancement of sustainable practices [24, 49]. A notable conflict arises between local communities and environmental advocates

African local people often depend on natural resources such as land, water, and forests for their livelihood. These resources are essential for agriculture, aquaculture, hunting, and foraging. The use of these resources by local people may lead to environmental degradation, deforestation, and a reduction in biodiversity. Conservationists assert that sustainable practices are crucial for environmental preservation and the enduring vitality of natural resources.

A fundamental dispute between local communities and environmental advocates concerns land rights. Conservationists often advocate for the creation of protected areas and wildlife reserves to save biodiversity and conserve endangered species. Nonetheless, these protected areas may restrict local residents' access to natural resources, leading to conflicts on land use and cultural practices.

Another source of conflict is the exploitation of natural resources for economic progress. Local communities may rely on industries such as forestry, mining, and agriculture for income and family sustenance. Conservationists argue that these activities might negatively impact the environment, leading to deforestation, pollution, and habitat destruction.

Moreover, conflicts arise from differing cultural values and goals. Local communities may prioritise immediate economic demands and traditional practices, while environmentalists focus on long-term ecological sustainability and conservation efforts. These divergent objectives lead to misunderstandings and conflicts between the two parties.

- **Solution**

To reconcile these conflicting interests, it is essential to include local communities in decision-making and conservation initiatives. Through collaboration with local stakeholders and respect for their traditional knowledge and customs, conservationists can create sustainable solutions that benefit both the environment and local communities.

Another approach is to promote community-based conservation projects that empower local communities to manage and protect their natural resources. Involving local populations in conservation efforts can promote sustainable development while respecting traditional customs and cultural values.

6.4.4 *Consumers Versus Producers*

In Africa, environmental stakeholders display conflicting interests about environmental conservation and the advancement of sustainable practices. A fundamental conflict arises between consumers and producers, who often have opposing interests and perspectives about the achievement of environmental sustainability [35, 50].

African consumers are becoming aware of the environmental ramifications of their purchase habits and are championing more sustainable products and practices. They are concerned about issues such as deforestation, water pollution, and climate change, and are likely to support companies that prioritise environmental sustainability. Consumers are more vocal in their demands for corporate responsibility regarding environmental policy, participating in actions such as boycotting products and advocating for legislative improvements.

Conversely, African producers face contradictory pressures to enhance profitability while meeting consumer demand for affordable products. A plethora of producers operate in industries with significant environmental impact, such as mining, agriculture, and manufacturing. These industries often rely on ecologically harmful practices, such as deforestation, water pollution, and greenhouse gas emissions. Producers tend to resist change due to concerns about increased costs, reduced profitability, or potential loss of market share.

- **Solution**

The conflicting interests of consumers and producers in Africa highlight the need for more collaboration and involvement among stakeholders. Consumers may significantly affect the demand for sustainable products and urge producers to adopt more environmentally responsible practices. By supporting companies that prioritise sustainability and advocating for stricter environmental regulations, consumers may cultivate a market for sustainable products and incentivise manufacturers to adopt enhancements.

In addition, producers may get benefits by aligning their business operations with consumer demand for sustainability. Employing more sustainable practices, firms may improve their brand, broaden their customer base, and mitigate regulatory issues. Producers may find that investing in sustainability yields cost savings, improves efficiency, and fosters long-term corporate resilience.

6.4.5 *Scientists Versus Sceptics*

There is an ongoing conflict among environmental stakeholders about environmental preservation and the advancement of sustainable practices in Africa. A fundamental difference arises between scientists, who emphasise the need of addressing environmental issues, and sceptics, who question the validity of climate change and other ecological problems [38, 51–53].

Scientists are crucial for understanding and addressing environmental challenges. They conduct research, assess data, and provide evidence-based recommendations to policymakers and the public. Researchers have warned that human actions, such as deforestation, urbanisation, and fossil fuel burning, are exacerbating global warming, increasing sea levels, and causing biodiversity loss. They assert that prompt actions are crucial to mitigate these impacts and provide a sustainable future for future generations.

Sceptics, on the other hand, question the scientific foundation of climate change and other ecological issues. They assert that natural occurrences, like solar activity and volcanic eruptions, substantially impact the Earth's climate system, while human activities have less influence. Sceptics often use the uncertainty inherent in climate models and past failed predictions to challenge the urgency of addressing environmental concerns.

The conflicting interests of scientists and sceptics significantly impact environmental policy and action. Scientists advocate for the implementation of sustainable practices, such as renewable energy sources, carbon pricing, and conservation strategies; yet, sceptics tend to resist these programmes, citing economic concerns or political ideologies. This gap hinders progress in addressing environmental issues and cause delays in implementing suitable remedies.

- **Solution**

To reconcile scientists and sceptics and foster collaboration on environmental issues, it is essential to promote open dialogue, transparency, and mutual understanding. Scientists may engage sceptics by addressing their concerns, presenting evidence-based arguments, and emphasising the importance of environmental protection initiatives. Sceptics may objectively analyse scientific data, consider the consequences of environmental degradation, and endeavour to reach an agreement on sustainable practices.

6.4.6 *Urban Versus Rural*

Conflicting interests often emerge among environmental stakeholders, particularly between urban and rural factions [40]. The main source of conflict between urban and rural interests is land use. Urban areas often face the challenge of limited space coupled with rising populations, necessitating land development for residential and infrastructure needs. In contrast, rural areas often emphasise agriculture and the preservation of natural ecosystems. This leads to conflicts around urban expansion, deforestation, and the intrusion of development into animal habitats.

A different sort of conflict emerges from conflicting objectives in resource management. Urban stakeholders tend to prioritise issues such as pollution and waste management, while rural stakeholders tend to focus on challenges like water conservation and land preservation. This sometimes leads to conflicts about the distribution and administration of resources, since urban and rural areas may have divergent goals related to resource use.

Furthermore, differences in environmental views and attitudes may be present between urban and rural stakeholders. Urban areas may emphasise prominent environmental initiatives, such as green spaces and renewable energy projects, while rural communities may prioritise traditional practices and land stewardship. This may hinder the establishment of agreement on optimum measures for environmental conservation and the advancement of sustainable practices.

- **Solution**

Despite these challenges, opportunities are available for urban and rural stakeholders to cooperate in discovering solutions that benefit both parties and the environment jointly. Collaboration and communication among diverse stakeholders may promote the resolution of conflicts and achieve consensus on issues of shared interest. Cooperative initiatives among urban and rural stakeholders can foster sustainable practices that protect the environment while meeting community needs.

6.5 Conclusion

Stakeholder engagement is a crucial aspect of environmental communication, involving individuals, groups, or organisations with a vested interest or affected by environmental issues. This includes government agencies, non-profit organisations, scientists, industry groups, consumers, and community members. Each stakeholder group brings unique perspectives and values, making their engagement crucial for effective communication and implementation of environmental initiatives. Building relationships with stakeholders, fostering trust, and promoting transparency is essential for building credibility, increasing buy-in for environmental projects, and driving positive change. Therefore, stakeholders play a vital role in environmental communication and organisations must prioritise building relationships and involving them in decision-making processes to effectively address environmental challenges and address their conflicts and potential solutions to these conflicts.

Recap

Stakeholder management plays a vital role in the operations of any organisation, particularly in the realm of environmental communication in Africa. Given the continent's diverse ecological challenges and socioeconomic issues, effective stakeholder engagement is imperative. By identifying, assessing, and engaging with a diverse range of stakeholders, organisations can enhance their credibility and effectiveness. This engagement not only fosters support for environmental initiatives but also empowers local communities, ensures their voices are heard, and aids in addressing the pressing environmental challenges facing the continent. In this way, sustainable development efforts can be more inclusive, effective, and resilient.

Questions

- What is stakeholder engagement in environmental communication?
- What is the importance of stakeholder engagement in environmental communication?
- What are the categories of stakeholders in environmental communication?
- What are the disputes among stakeholders and potential solutions to these conflicts?

References

1. BMU. Federal Ministry for the Environment, Nature Conservation and Nuclear Safety, Division WR II 3: annual report on end-of-life vehicle reuse/recycle/recovery rates in Germany for 2015; 2017. http://www.bmu.de/fileadmin/Daten_BMU/Download_PDF/Abfallwirtschaft/jahresbericht_altfahrzeug_2015_en_bf.pdf.
2. BMU. Federal Ministry for the Environment, Nature Conservation and Nuclear Safety, Division G I 5: GreenTech made in Germany—environmental Technology Atlas for Germany; 2018. https://www.bmu.de/fileadmin/Daten_BMU/Pools/Broschueren/greentech_2018_en_bf.pdf.
3. Cahill R, Grimes SM, Wilson DC. Extended producer responsibility for packaging wastes and WEEE – a comparison of implementation and the role of local authorities across Europe. WM&R. 2011;29(5):455–79. https://doi.org/10.1177/0734242X10379455.
4. Andrew CP. The evolution of strategic corporate social responsibility. EuroMed J Bus. 2013;8(3):220–42. https://doi.org/10.1108/EMJB-05-2013-0030.
5. CENN. Waste recycling potential (in Georgia). Caucasus Environmental NGO Network CENN; 2016. http://environment.cenn.org/app/uploads/2016/09/waste_eng.pdf. Accessed Mar 2018.
6. Dace E, Pakere I, Blumberga D. Evaluation of economic aspects of the deposit-refund system for packaging in Latvia. EQM. 2013 Apr;24(3):311–29. https://doi.org/10.1108/14777831311322631.
7. Díaz-García C, González-Moreno Á, Sáez-Martínez FJ. Eco-innovation: insights from a literature review. Innovations. 2015;17(1):6–23. https://doi.org/10.1080/14479338.2015.1011060.
8. DSD. DSD—Duales system holding GmbH & Co. KG. Sustainability report 2015/2016; 2017. https://www.gruener-punkt.de/fileadmin/layout/redaktion/Nachhaltigkeit/Nachhaltigkeitsbericht_15-16/ENG/DSD_NHB_ENG_Komplett_WEB_Final.pdf.
9. EC European Commission DG ENV Plastic waste: ecological and human health impacts, *In-Depth Report*, November 2011; 2011. http://ec.europa.eu/environment/integration/research/newsalert/pdf/IR1_en.pdf.
10. Amor-Esteban V, Galindo-Villardón MP, García-Sánchez IM. Useful information for stakeholder engagement: a multivariate proposal of an Industrial Corporate Social Responsibility Practices Index. Sustain Dev. 2018;26(6):620–37. https://doi.org/10.1002/sd.1732.
11. Barr S, Guilbert S, Metcalfe A, Riley M, Robinson GM, Tudor TL. Beyond recycling: an integrated approach for understanding municipal waste management. Appl Geogr. 2013;39:67–77. https://doi.org/10.1016/j.apgeog.2012.11.006.
12. BIO. BIO Intelligence Service: review of the scope of the Directive 2012/19/EU on Waste Electrical and Electronic Equipment (WEEE). Final report prepared for the European Commission—DG Environment; 2013. http://ec.europa.eu/environment/waste/weee/pdf/approved_%20WEEE%20scope%20review.pdf (Cited June 2016).
13. BIO. BIO Intelligence Service: development of guidance on extended producer responsibility (EPR). Report for the European Commission DG ENV 2017. http://ec.europa.eu/environment/waste/pdf/target_review/Guidance%20on%20EPR%20-%20Final%20Report.pdf.

14. Birch A, Hon KK, Short T. Structure and output mechanisms in Design for Environment (DfE) tools. J Clean Prod. 2012;35:50–8. https://doi.org/10.1016/j.jclepro.2012.05.029.
15. EC. Directive 2012/19/EU of the European parliament and of the commission on WEEE; 2012. https://eur-lex.europa.eu/legal-content/EN/TXT/PDF/?uri=CELEX:32012L0019&from=EN.
16. El Akremi A, Gond JP, Swaen V, De Roeck K, Igalens J. How do employees perceive corporate responsibility? Development and validation of a multidimensional corporate stakeholder responsibility scale. JOM. 2018;44(2):619–57. https://doi.org/10.1177/0149206315569311.
17. Freeman RE. The politics of stakeholder theory: some future directions. In: R. Edward Freeman's selected works on stakeholder theory and business ethics. Cham: Springer International Publishing; 2023. p. 119–32. https://doi.org/10.5840/10.2307/3857340.
18. Gallardo A, Bovea MD, Colomer FJ, Prades M, Carlos M. Comparison of different collection systems for sorted household waste in Spain. WM. 2010;30(12):2430–9. https://doi.org/10.1016/j.wasman.2010.05.026.
19. Gallardo A, Carlos M, Colomer FJ, Edo-Alcón N. Analysis of the waste selective collection at drop-off systems: case study including the income level and the seasonal variation. WM&R. 2018;36(1):30–8. https://doi.org/10.1177/0734242X17733539.
20. Gupt Y, Sahay S. Review of extended producer responsibility: a case study approach. WM&R. 2015;33(7):595–611.
21. GVM. Reusable and ecologically advantageous one-way packaging quota 215. Gesellschaft für Verpackungsmarktforschung mbH; 2017. http://gvmonline.de/files/blickpunkt/2017_06_EWMW2015_en.pdf.
22. Huisman J, Botezatu I, Herreras L, et al. Countering WEEE illegal trade (CWIT) summary report, market assessment, legal analysis, crime analysis and recommendations roadmap. Lyon, France; 2015, p. 61. http://www.cwitproject.eu/wp-content/uploads/2015/09/CWIT-Final-Report.pdf.
23. Janz A, Bilitewski B. WEEE in and outside Europe-hazards, challenges and limits. In: Prosperity waste and waste resources. Proceedings of the 3rd BOKU waste conference. Vienna: BOKU-University of Natural Resources and Applied Life Sciences; 2009. p. 113–22.
24. Kazadi K, Lievens A, Mahr D. Stakeholder co-creation during the innovation process: identifying capabilities for knowledge creation among multiple stakeholders. J Bus Res. 2016;69(2):525–40. https://doi.org/10.1016/j.jbusres.2015.05.009.
25. Li J, Liu L, Zhao N, Yu K, Zheng L. Regional or global WEEE recycling. Where to go? WM. 2013;33(4):923–34. https://doi.org/10.1016/j.wasman.2012.11.011.
26. Lidl Positionspapier zur Pfandpflicht für PET-Einwegflaschen in Deutschland. [In German]; 2017. https://www.lidl.de/de/asset/other/20160429_Positionspapier-Pfand.pdf.
27. Markovic S, Bagherzadeh M. How does breadth of external stakeholder co-creation influence innovation performance? Analyzing the mediating roles of knowledge sharing and product innovation. J Bus Res. 2018;88:173–86. https://doi.org/10.1016/j.jbusres.2018.03.028.
28. Marshall RE, Farahbakhsh K. Systems approaches to integrated solid waste management in developing countries. WM. 2013;33(4):988–1003. https://doi.org/10.1016/j.wasman.2012.12.0123.

29. NABU. Das Geschäft mit dem Einwegpfand. NABU-Hintergrundpapier. [In German]; 2017. https://www.nabu.de/imperia/md/content/nabude/abfallpolitik/170207_nabu_infopapier_einwegpfand.pdf.
30. OECD. Extended producer responsibility: a guidance manual for governments. Paris: OECD; 2021. https://doi.org/10.1787/9789264189867-en.
31. OECD. Stakeholder engagement. In: Government at a glance. Paris: OECD; 2015. https://doi.org/10.1787/gov_glance-2015-38-en.
32. OECD. Extended producer responsibility: updated guidance for efficient waste management. Paris: OECD; 2016. https://doi.org/10.1787/9789264256385-en.
33. Oguchi M, Sakanakura H, Terazono A, Takigami H. Fate of metals contained in waste electrical and electronic equipment in a municipal waste treatment process. WM. 2012;32(1):96–103. https://doi.org/10.1016/j.wasman.2011.09.012.
34. Ongondo FO, Williams ID, Cherrett TJ. How are WEEE doing? A global review of the management of electrical and electronic wastes. WM. 2011;31(4):714–30. https://doi.org/10.1016/j.wasman.2010.10.023.
35. Pires A, Martinho G, Chang NB. Solid waste management in European countries: a review of systems analysis techniques. J Environ Manag. 2011;92(4):1033–50. https://doi.org/10.1016/j.jenvman.2010.11.024.
36. Quan Y, Wu H, Li S, Ying SX. Firm sustainable development and stakeholder engagement: the role of government support. BSE. 2018;27(8):1145–58. https://doi.org/10.1002/bse.2057.
37. Reed MS, Vella S, Challies E, et al. A theory of participation: what makes stakeholder and public engagement in environmental management work? Restor Ecol. 2018;26:S7–17. https://doi.org/10.1111/rec.12541.
38. Ridder HG. The theory contribution of case study research designs. Bus Res. 2017;10:281–305. https://doi.org/10.1007/s40685-017-0045-z.
39. Salhofer S, Obersteiner G, Schneider F, Lebersorger S. Potentials for the prevention of municipal solid waste. WM. 2008;28(2):245–59. https://doi.org/10.1016/j.wasman.2007.02.026.
40. Schnoor JL. Extended producer responsibility for e-waste. Environ Sci Technol. 2012;46(15):7927. https://doi.org/10.1021/es302070w.
41. Shams SR. Transnational education and total quality management: a stakeholder-centred model. J Manag Dev. 2017;36(3):376–89. https://doi.org/10.1108/JMD-10-2015-0147.
42. Riad Shams SM, Belyaeva Z. Quality assurance driving factors as antecedents of knowledge management: a stakeholder-focussed perspective in higher education. J Knowl Econ. 2019;10:423–36. https://doi.org/10.1007/s13132-0170472-2.
43. Shams SR, Kaufmann HR. Entrepreneurial co-creation: a research vision to be materialised. Mgmt Decis. 2016;54(6):1250–68. https://doi.org/10.1108/MD-11-2015-0487.
44. Taylor DC. Policy incentives to minimize generation of municipal solid waste. WM&R. 2000;18(5):406–19. https://doi.org/10.1034/j.1399-3070.2000.00156.
45. Tencati A, Pogutz S, Moda B, Brambilla M, Cacia C. Prevention policies addressing packaging and packaging waste: some emerging trends. WM. 2016;56:35–45. https://doi.org/10.1016/j.wasman.2016.06.025.
46. UBA how to do ecodesign? Result of a research project commissioned by the German Federal Environment Agency; 2015. https://www.umweltbundesamt.de/sites/default/files/medien/376/publikationen/ecodesign_eng_14_3.epub.

47. UBA Bundesweite Erhebung von Daten zum Verbrauch von Getränken [in German]; 2017.
48. Ahmed M. The role of indigenous communication systems in environmental conservation: a case study of the Maasai Community in Kenya. J Afr Cult Stud. 2017;29(3):325–40.
49. Bello R. Harnessing the power of folklore in environmental communication: insights from West Africa. J Environ Comm. 2019;12(2):187–201.
50. Germany Packaging Act, July 2017. [In German]; 2017. https://www.bgbl.de/xaver/bgbl/start.xav?start=//*%5b@attr_id=%27%27%5d#__bgbl__%2F%2F*%5B%40attr_id%3D%27bgbl117s2234.pdf%27%5D__1535444822799.
51. Gonzalez-Urango H, García-Melón M. Stakeholder engagement to evaluate tourist development plans with a sustainable approach. SD. 2018;26(6):800–11. https://doi.org/10.1002/sd.1849.
52. Guerrero LA, Maas G, Hogland W. Solid waste management challenges for cities in developing countries. WM. 2013;33(1):220–32. https://doi.org/10.1016/j.wasman.2012.09.008.
53. Gui L, Atasu A, Ergun Ö, Toktay LB. Efficient implementation of collective extended producer responsibility legislation. Manag Sci. 2016;62(4):1098–123. https://doi.org/10.1287/mnsc.2015.2163.

CHAPTER 7

The Role of Environmental Communication in Education

Introduction

The importance of environmental education in today's society cannot be overemphasised. Environmental communication programmes have an obligation to educate the current generation and future generations about the need of maintaining and safeguarding the environment in light of the numerous ecological issues threatening it, including pollution, deforestation, and climate change. Environmental communication is essential to this education process because it raises awareness of environmental issues and motivates people to take action. The primary focus of this chapter is the function of environmental communication in education. The first part of the chapter looks at the role that environmental communication plays in raising public awareness of environmental issues. The chapter then discusses how environmental communication shapes academic curricula by raising awareness about environmental issues and promoting sustainability. The chapter ends by highlighting some of the successful initiatives in integrating environmental communication in education.

7.1 The Importance of Environmental Communication in Education

One of the primary purposes of environmental communication in the educational system is to raise public awareness of environmental issues [1]. There are many environmental issues that require immediate attention, such as deforestation, pollution, and biodiversity loss [1]. By providing tools and information,

© The Author(s), under exclusive license to Springer Nature Switzerland AG 2025
G. Mutinta, *An African Perspective on Environmental Communication*, https://doi.org/10.1007/978-3-032-05052-6_7

environmental communication helps people comprehend the causes, consequences, and potential solutions of numerous situations.

In addition to raising awareness, environmental communication is crucial for fostering a sense of responsibility among individuals. Communication helps us comprehend our role as stewards of the planet by highlighting the interconnection of all living things and the natural environment [2]. It challenges people to consider the consequences of our actions and choices on the environment and the coming generation. By promoting a sense of responsibility, environmental communication empowers individuals to take control of their lives and create positive changes.

Furthermore, by providing opportunities for participation in environmental preservation activities, environmental communication inspires individuals to take action. Through activism, volunteer work, or lifestyle modification, communication provides a means for individuals to become engaged in environmental issues and make a meaningful difference in the world [2]. Through the presentation of innovative concepts, real-world examples, and useful resources, environmental communication motivates individuals to take action in the direction of a more sustainable future.

Environmental communication, as a whole, is an important educational technique that raises awareness of the need of environmental conservation and inspires people to take action [3]. By raising awareness, fostering responsibility, and inspiring action, communication enables individuals to become informed, engaged citizens who have the capacity to improve the environment. Effective communication can educate and motivate people to protect and save our planet for future generations.

7.1.1 Advocacy and Activism

Environmental communication can help to educate the public about environmental issues and the importance of taking action [4]. Understanding the science behind environmental threats, and the potential consequences if left unchecked, can help people appreciate their seriousness. As a result of this awareness, people might be motivated to engage in environmental activism and advocacy through signing petitions, attending demonstrations, or making charitable contributions to environmental causes.

Environmental communication can inspire people to take action on environmental issues by presenting success stories and the advantages that can be realised through teamwork [5]. By sharing stories of local communities coming together to support renewable energy sources, plant trees, or clean up beaches, environmental communicators can encourage others to take action and get involved in environmental protection. These victories can serve as powerful examples of the beneficial effects that result from people banding together in support of a common objective.

Additionally, environmental communication can facilitate cooperation and dialogue amongst various stakeholders, including communities, businesses, environmental organisations, and policymakers [6]. Environmental communicators can assist narrow gaps between diverse groups and form alliances to effect meaningful change on environmental challenges by promoting open and positive communication. Stakeholders can address serious environmental concerns by communicating and collaborating to identify solutions, implement legislation, and gather resources.

Moreover, environmental communication can give people the means to demand legislative changes and hold decision-makers accountable for their environmental actions [7]. By providing people with the knowledge and tools they need to become advocates, environmental communication can aid in increasing public awareness of issues related to environmental protection. People can organise advocacy campaigns, write letters to legislators, or testify at public hearings to ensure that their voices are heard and that action is taken on environmental issues.

7.1.2 Public Relations and Marketing

Environmental communication can help raise public awareness of the importance of sustainable practices and environmental conservation [8]. Businesses and organisations can draw consumers' attention to pressing environmental issues in order to encourage them to make environmentally friendly decisions. Also, this can alert customers to the necessity of taking action. With targeted messaging and campaigns, businesses can become the sustainability leaders in their industry and inspire others to follow suit.

With the public and stakeholders, one can build credibility and trust through environmental communication. By being transparent and truthful about their environmental initiatives, businesses can demonstrate their commitment to moral business conduct and accountability for their environmental impact. Enhancing the company's reputation and image can lead to a rise in customer loyalty and trust.

Besides, environmental communication can also help a business stand out from the competition and attract environmentally conscious clients. In today's competitive economy, customers are searching for businesses that prioritise environmental stewardship and sustainability [8]. By effectively communicating about their green initiatives and showcasing their commitment to environmental responsibility, businesses can set themselves apart and draw in a growing segment of environmentally conscious consumers.

Environmental communication has a lot to offer through advocacy and social change [9]. By maximising their platform and resources, businesses and organisations can use their voice to influence public opinion, influence policy decisions, and promote positive environmental outcomes. Through strategic communication efforts, they can mobilise support for environmental causes, engage with key stakeholders, and form relationships with like-minded organisations in order to effect meaningful change.

7.1.3 Community Engagement

One way of increasing community awareness of environmental issues is through environmental communication [10]. Environmental communication can assist community members in comprehending the significance of these issues and their impact on the environment by furnishing them with succinct and comprehensible information. This information enables individuals to make more informed decisions and effectively address these issues in their personal lives.

The capacity to motivate community action is a characteristic of environmental communication [11]. Environmental communication can inspire individuals to become involved and make a positive impact on the world by emphasising notable initiatives, success tales, and the benefits of environmental endeavours. Environmental communication has the potential to motivate individuals in the community to advocate for policy changes, support sustainable businesses, and address environmental concerns.

Additionally, environmental communication is a valuable instrument for cultivating collaboration and relationships within the community [12]. Environmental communication can foster debate and idea exchange among stakeholders, organisations, and community members by establishing transparent and accessible communication channels. This collaborative approach has the potential to foster harmony and shared accountability for environmental issues when combined with more practical solutions.

In order to resolve urgent environmental challenges, environmental communication can be employed to instil a sense of urgency in the community [13]. Environmental communication can motivate community members to take immediate action to safeguard the environment by highlighting the pressing issues that threaten the planet, including water scarcity, air pollution, and deforestation. Environmental communication can inspire individuals to prioritise environmental concerns and strive for meaningful change by depicting environmental issues as imperative and vital.

7.1.4 Political Advocacy

Environmental communication can help raise public awareness of and educate people about the main environmental issues that Africa is currently facing [14]. A range of media channels, such as social media, news outlets, and community events, can be used to spread awareness of issues such as biodiversity loss, air and water pollution, deforestation, and climate change. With the help of this awareness, public support can be mobilised for laws and policies that make an effort to address these problems.

Environmental communication can arouse people's sense of urgency and inspire them to take action. When communicators highlight the immediate and long-term consequences of environmental problems on ecosystems, wildlife, and human health, they can inspire people to support change. By highlighting the adverse consequences of climate change, such as extreme weather, rising sea levels, and food insecurity, legislation aimed at reducing greenhouse gas emissions and promoting renewable energy sources, for example, can be passed into law.

Furthermore, environmental communication can facilitate cooperation and communication amongst stakeholders holding disparate interests and points of view [15]. By encouraging open, honest, and inclusive discussions about environmental issues, individuals can reach an understanding, identify common goals, and develop innovative solutions. For example, policymakers, scientists, environmentalists, businesses, and community members can collaborate to discuss sustainable development and develop policies that balance economic growth with environmental protection.

Environmental communication can help hold policymakers and corporations accountable for their decisions and deeds [16]. Investigative reporting, public demonstrations, and advocacy campaigns can be used to reveal environmental infractions, corruption, and carelessness. Decision-makers can be under pressure to prioritise environmental preservation and enforce environmental laws as a result. Holding companies responsible for deforestation or pollution of waterways, for example, can result in fines, legal action, and changes to corporate policies.

7.2 Storytelling

Environmental communication through storytelling is essential to education because it makes difficult environmental topics more approachable and interesting for learners. Using stories, visuals, and other storytelling techniques, environmental communication can create a more immersive learning environment that inspires people to take action and change the world around them:

- **Telling personal stories:** One of the ways to communicate environmental issues is through the use of educational personal stories [16]. By sharing personal accounts of how environmental issues have affected them, communicators can establish a connection and evoke feelings in their audience. The story of a family that faces difficulties in accessing clean water due to pollution, for example, serves to highlight the pressing need for environmental action.
- **Visual storytelling:** People can communicate complex environmental issues in an engaging way by using visual storytelling [17]. Using visuals like images, videos, and infographics, communicators can convey information to their audience quickly and effectively while also evoking feelings in them. A movie that illustrates the impact of deforestation on wildlife, for example, can highlight how important it is to protect our forests.
- **Collaborative storytelling:** A range of stakeholders, including experts, activists, and members of the community, can be involved in collaborative storytelling to create an extensive narrative on environmental issues. By

bringing together different viewpoints and voices, communicators can provide an integrated perspective on the issues people face and the solutions available to address them. This tactic can boost confidence in individuals to take environmental action and promote cooperation.

- **Storytelling with a focus on solutions:** While raising awareness of environmental problems is important, it is also critical to draw attention to potential solutions and encourage action. Solutions-oriented storytelling can draw attention to innovative ideas, successful campaigns, and positive outcomes in the environmental field. By sharing success stories and providing them with the resources they need to make a difference, environmental communication professionals can motivate and inspire their audience to take action and create a more sustainable future.

7.3 Collaboration

Environmental communication facilitates public education and awareness-building regarding important environmental issues [18]. Effective communication of the consequences of human activity on the environment can increase the likelihood that people will acknowledge the need for action and change their behaviour to leave a smaller environmental footprint. When people are more aware of environmental issues, they are more likely to come together to discuss them and work towards finding solutions.

Among the stakeholders that environmental communication can facilitate engagement and communication with are governments, corporations, non-profit organisations, and communities [19]. Creating a space for discussion and the sharing of ideas can encourage collaboration in the development of comprehensive and effective solutions for environmental problems. By encouraging open communication and active participation, stakeholders can establish shared objectives and priorities and create strategies for addressing them together.

Environmental communication can facilitate the formation of alliances and partnerships between various groups who share a common interest in environmental sustainability and conservation. By promoting collaboration and teamwork, environmental communication can unite individuals and groups with disparate origins, resources, and perspectives in order to work towards common goals. If collaborative efforts leverage the strengths and capabilities of each stakeholder, they can have a greater chance of producing positive environmental outcomes.

Through environmental communication, people and communities can be given the power to take control of their own local environments and bring about change [19]. If people have access to resources, information, and communication tools, they can become more informed and involved in environmental issues. This can lead to community-based efforts and projects that have a significant impact on the environment. Encouraging people to become environmental stewards can serve as a catalyst for grassroots collaboration and positive change.

7.4 Building Resilience to Climate Change

Environmental communication can help people understand the causes and implications of climate change [19]. By disseminating correct scientific information and raising awareness of the need to reduce greenhouse gas emissions, environmental communication can enable individuals and communities to take action to minimise their carbon footprint and adapt to climate change. Environmental communication may educate people about the urgency of the climate situation and inspire them to take action to protect the environment for future generations by implementing outreach activities and focused messages.

Besides, environmental communication can serve to unite and strengthen stakeholders such as governments, businesses, non-profit organisations, and the general public [20]. People may collaborate to develop innovative climate change solutions and increase community resilience by having open and honest conversations. Effective communication techniques can encourage resource sharing, collaboration, and the adoption of sustainable behaviours that can help to alleviate the consequences of climate change and enhance long-term resilience.

Through environmental communication, the public can be educated and mobilised to support policies and activities that promote sustainability and resilience. By framing climate change as an important issue affecting everyone, environmental communication can raise awareness of the problem and motivate collaboration to address its core causes. Communication may engage the public in meaningful discussions on climate change through media campaigns, social media outreach, and public events, as well as encourage them to support policies that promote renewable energy, conservation, and sustainability. By leveraging the power of communication, we can build a broad base of support for climate action and create a more resilient society.

Environmental communication can also help to create credibility and trust in the scientific community, as well as support evidence-based decision-making [21]. By speaking openly and honestly about the risks and uncertainties associated with climate change, campaigns can help debunk myths and incorrect information that damage public trust in science. Campaigns that use strong messaging and narrative can help the general public comprehend the difficult science behind climate change and encourage people to act. Environmental communication may help people, organisations, and politicians make informed decisions that promote sustainability and resilience in the face of climate change by establishing a solid foundation of credibility and confidence.

7.5 Promoting Environmental Justice

Environmental communication can be used as an educational tool for environmental justice, raising awareness of environmental challenges that disproportionately affect marginalised people [22]. Environmental communication can raise awareness of the need for action and demand justice for those most affected by pollution, climate change, and environmental degradation by highlighting the ways in which low-income communities, communities of colour, and indigenous communities are disproportionately impacted.

In addition to promoting awareness, environmental communication may give marginalised people the tools they need to advocate for their own rights and interests. Environmental communication may help communities organise, mobilise, and support practices and policies that promote environmental justice while also protecting their health and well-being.

Businesses and governments can be held accountable for acts and policies that promote environmental injustice through environmental communication [23]. By uncovering and publicising incidents of environmental racism, exploitation, and neglect, environmental communication can put pressure on legislators to accept responsibility for their actions and implement laws that promote justice and equity for all citizens.

Environmental communication can also foster collaboration and solidarity among many groups and stakeholders demanding environmental justice. Encouraging communication, sharing resources and expertise, and establishing ties across different groups and ideas can all help our organisation's efforts to create a fairer and sustainable environment for future generations.

7.6 Mobilising Support for Environmental Initiatives

Public awareness of urgent environmental issues can be increased through environmental communication [24]. People's awareness of the seriousness of environmental issues such as pollution, deforestation, climate change, and others can be increased through concise and straightforward science-based communication. Environmental communicators can engage and educate the public about the significance of taking action to protect the planet by using powerful stories, powerful imagery, and simple language.

Environmental communication can drive people to take action. Communicators can emphasise the positive effects that individual acts can have on the environment in order to urge individuals to make more sustainable decisions in their daily lives. Environmental communication can assist people understand the importance of their actions in combating environmental degradation and climate change by encouraging them to minimise waste, preserve energy, or support eco-friendly enterprises [23].

Those who are enthusiastic about environmental issues also feel more united and supportive of one another when environmental communication is promoted [24]. Environmental communicators can facilitate connections between individuals and organisations that share similar values by creating forums for discussion, cooperation, and dialogue. When people band together to support environmental causes, push for legislative changes, and hold governments and corporations responsible for their environmental effects, collective action can be ignited.

Furthermore, environmental communication can enable people to take on leadership roles in relation to environmental issues in their local communities [25]. Environmental communicators can assist people in becoming environmental stewards by giving them the information, instruments, and resources they require to take action. Environmental communication can assist individuals in taking proactive steps to enhance their local environment, like establishing a community garden, planning a beach clean-up, or getting in touch with lawmakers.

7.7 Integrating Environmental Communication into Educational Curricula

Environmental communication is essential to teaching the next African generation the importance of protecting the environment. By strengthening learners' grasp of environmental issues and integrating environmental communication into curricula, learners can be equipped to take action.

7.7.1 Raising Awareness

Curriculum that includes environmental communication can help learners better understand the importance of sustainability and the urgency of environmental issues. If learners are exposed to modern environmental issues, they are more likely to develop an environmental responsibility. Learning environmental communication improves learners' ability to make environmentally friendly decisions. By introducing environmental communication into several subjects, learners can gain a thorough understanding of environmental issues and how they affect society [26].

Integrating environmental communication into the curriculum can promote problem-solving and critical thinking in the learners [26]. When learners participate in discussions about environmental issues, they are motivated to think critically about challenging environmental problems and develop original solutions. By looking at different viewpoints and recognising how interconnected environmental issues are, learners can develop a deeper understanding of the environment and its significance. Through environmental communication, learners can collaborate with one another, take part in group discussions, and engage in hands-on activities that promote environmental awareness and action.

By introducing environmental communication into the curriculum, learners can gain the information and skills needed to actively effect change in their communities [27]. Learners who understand how their actions affect the environment can make more environmentally conscious decisions. Environmental communication can motivate learners to engage in eco-friendly activities such as recycling, waste reduction, and energy conservation. Integrating environmental communication into topics like language arts, media studies, and ethics can help students express themselves clearly and urge others to act. Through project-based learning and community engagement programmes, students can apply their knowledge and abilities to address local environmental concerns and make a positive difference in Africa.

7.7.2 *Promoting a Sense of Responsibility*

Educating learners in the classroom about environmental communication can help them understand the links between environmental issues and the consequences of human activity on the environment. Studying topics like deforestation, pollution, and climate change can help learners gain a deeper understanding of the complex issues facing the environment and their role in resolving them. By being better prepared to make decisions about their own behaviour and lifestyle, learners can demonstrate a stronger commitment to sustainability and make more responsible decisions.

Including environmental communication into curricula can help learners develop their critical thinking skills and their ability to evaluate and analyse environmental information. By interacting with a variety of informational materials, such as scientific studies, news articles, and documentaries, learners can develop critical thinking skills, the ability to distinguish between fact and opinion, and the ability to assess the reliability of sources. Learners with these critical thinking skills can develop into more informed and responsible adults who are able to lobby for positive community change and make well-informed decisions about environmental matters.

Integrating environmental communication into curricula can help learners develop empathy and compassion for the natural world and all living things [28]. Understanding the interdependence of ecosystems, the importance of biodiversity, and the impacts of human activity on wildlife and habitats can help learners have a greater appreciation for the diversity and beauty of the natural world. Learners with this empathy could be motivated to support conservation efforts, take up environmental protection, and make sustainable decisions in their daily lives.

7.7.3 *Encouraging Critical Thinking*

More than ever, it important to instil in the next generation of Africans the importance of critical thinking abilities through environmental communication. The continent is rapidly evolving [14]. The value of environmental communication in promoting critical thinking cannot be overstated, since it not only increases people's awareness of their surroundings but also pushes them to think about how their actions impact the environment.

- **Encourages multidisciplinary thinking:** Learners need to consider a variety of perspectives and solutions because environmental issues are complex and multifaceted. By incorporating environmental communication into different courses, educators can encourage learners to think across disciplines. By learning to draw connections between ideas from diverse academic fields, learners can hone their critical thinking skills and prepare for environmental challenges in the real world [15].

- **Strengthens analytical skills:** Environmental communication involves analysing data, evaluating supporting information, and identifying patterns and trends. When environmental topics are incorporated into curricula, learners can be exposed to real-world environmental issues and given the chance to critically analyse information. For example, learners could investigate how climate change affects a local ecosystem or evaluate the efficacy of environmental policies. These experiences, which also aid in the development of analytical skills, can inspire learners to consider the causes and effects of environmental problems critically.
- **Promotes the growth of problem-solving ability:** Environmental communication comprises discussing and coming up with solutions for environmental issues. By incorporating environmental themes into school curricula, educators can support learners' growth as problem-solvers. Learners can work on projects that involve project-based learning, such as researching renewable energy sources or developing sustainable communities. Learners can be able to improve their critical thinking abilities and problem-solving abilities by means of these hands-on experiences with potential solutions to environmental challenges [16].

7.7.4 Promoting Advocacy

The public's increased awareness of environmental issues is the primary advantage of incorporating environmental communication into curriculum development [17]. By instilling in learners the importance of environmental protection, institutions can inspire them to become change agents. When learners are taught about the negative effects of environmental problems, they are more likely to take action to protect the environment. Learners who are familiar with the issues can mentor others and hold thought-provoking debates regarding potential fixes.

Including environmental communication in the curriculum can help learners' critical thinking and communication skills develop. By studying environmental issues and honing their communication skills, learners can become more effective change agents. Learners can practise persuading others with their ideas through research papers, presentations, and projects related to environmental issues. These skills are essential for effective advocacy, since speaking clearly and persuasively is a prerequisite for both raising awareness and inspiring people to take action.

In addition, teaching environmental communication in the classroom can cultivate learners' empathy and sense of responsibility for the environment [29]. By studying the interdependence of all living things and the consequences of human activity on the environment, learners can develop a greater sense of empathy for the natural world. This empathy can encourage learners to advocate for sustainable practices and environmental protection. Learners who are instilled with a sense of environmental responsibility can become a generation of fervent supporters of constructive change.

7.7.5 *Building Communication Skills*

To meet the standards of environmental communication, learners must first learn how to convey complicated scientific information in an effective and understandable manner [2]. Learners should have a solid understanding of the scientific principles underlying environmental issues and the ability to translate that knowledge into language that non-expert readers can comprehend if they hope to effectively communicate these issues to the general public. By working on projects and assignments related to environmental communication, learners can practise condensing complex information into easily understood messages and improve their ability to present information in an engaging and compelling way.

Environmental communication requires interaction with a range of audiences and stakeholders, so learners can learn how to modify their message to fit different audiences [6]. Effective communication in the environmental sector often requires working with a diverse range of individuals, such as government officials, activists, and members of the community. By means of integrating environmental communication into academic curricula, learners can acquire proficiency in crafting messages that are persuasive and relevant to diverse audiences. They can also learn how to adjust their communication style to engage with different kinds of people and build relationships with key stakeholders.

Moreover, as learners engage with complex and varied environmental concerns, environmental communication can support the development of their critical thinking and problem-solving skills. When it comes to the causes and potential solutions of environmental problems, learners who wish to communicate about them effectively often need to engage in data analysis, evidence evaluation, and critical thinking. Learners can develop their critical and creative thinking abilities as well as their capacity to work with others to solve difficult problems through coursework that incorporates environmental issues from the real world.

7.7.6 *Promoting Creativity*

When faced with difficult problems, learners who are exposed to environmental communication in the classroom can be forced to think creatively and unconventionally [7]. Studying about issues like pollution, deforestation, and habitat loss forces learners to think of creative solutions in a sustainable and environmentally conscious manner. This can lead to the development of novel ideas and technological breakthroughs that help shape Africa in a more sustainable way in the future.

By integrating environmental communication into curricula, educators can inspire learners' natural world wonder and show curiosity. By learning about the intricate ecosystems that make up the planet and the delicate balance that exists between humans and the environment, learners can develop a deeper understanding of Africa they live in. This sense of wonder can inspire learners to research new ideas, ask questions, and seek out innovative ways to protect and preserve the environment.

Moreover, by incorporating environmental communication into academic curricula, learners can gain critical communication and teamwork skills—both essential for managing complex environmental issues [9]. Through group projects, discussions, and presentations on environmental issues, learners can work towards a common goal while honing their teamwork and communication skills. These skills are not just relevant in the context of environmental issues but also in a wide range of other fields and disciplines.

7.7.7 Connecting Classroom Learning to Environmental Issues in the Real World

Integrating environmental communication into curricula can facilitate learners' understanding of the connections between environmental issues and social, political, and economic factors [12]. Learners who embrace this comprehensive method of instruction can acquire a deep comprehension of the nuances of environmental problems and the various stakeholders involved. By studying case studies and engaging in discussions about real environmental issues, learners can acquire critical thinking skills and learn how to analyse and evaluate different points of view on environmental problems. This can enhance their comprehension of academic subjects and equip them to become informed, engaged citizens who can assist in resolving environmental issues.

Incorporating environmental communication into curricula can also encourage learners to apply their knowledge in practical settings [11]. By participating in practical projects such as organising environmental campaigns, conducting field research, or collaborating with local organisations, learners can observe first-hand the direct consequences of their actions on the environment and society. This experiential learning approach not only enhances learning in the classroom but also provides learners a sense of agency and empowers them to effect change in their communities. By drawing connections between their academic knowledge and pressing environmental issues, learners can be able to comprehend the relevance and significance of their education in addressing these pressing issues.

In addition, including environmental communication into curricula can encourage learners to acquire effective communication skills, which are essential for addressing environmental issues in a collaborative and inclusive manner. By engaging in debates, multimedia presentations, public speaking, report writing, and other relevant activities, learners can enhance their communication abilities to effectively explain intricate environmental matters to diverse audiences. Learners can use this to share their knowledge and viewpoints with others, advocate for environmental causes, and have meaningful conversations with stakeholders. Learners who work on their communication skills can be more adept at reaching agreements, having productive discussions, and mobilising others in favour of environmental action [9].

7.7.8 *Promoting Interdisciplinary Learning*

Incorporating environmental communication can encourage learners to think critically and creatively about environmental issues. To effectively explain complex environmental concepts to a variety of audiences, integrating environmental communication into learners' courses can help them apply their knowledge of science, technology, policy, and social sciences. This interdisciplinary approach can push learners to connect different fields of study and take an integrated approach to environmental challenges [11].

Integrating environmental communication into curricula can encourage learners to collaborate and work as a team to achieve common goals. Since environmental problems are inherently interdisciplinary, working solutions require experts from a variety of fields. By engaging in group projects and discussions centred on environmental communication, learners can develop the ability to work effectively with individuals from a variety of disciplines and backgrounds. Through this collaborative approach, learners can have a sense of shared responsibility and are encouraged to recognise the value of diverse perspectives [22].

Besides, integrating environmental communication into curricula can help learners acquire critical skills that they can utilise outside of the classroom. Learners who wish to communicate about the environment in an effective way need to be able to think critically, research topics thoroughly, and articulate their ideas clearly. Learners who put these abilities into practice can be more adept at communicating effectively in a variety of settings, including social media campaigns, public forums, and business presentations [29].

7.7.9 *Promoting Collaboration*

Collaboration is primarily promoted by integrating environmental communication into educational curricula because it instils in learners a sense of shared responsibility [9]. By studying environmental issues together, learners can make the connections between what they do and how their individual behaviour affects the world around them. This realisation can motivate learners to work together to solve environmental issues and develop a sense of collective responsibility. When learners collaborate on environmental projects or initiatives, they can combine their knowledge and resources to have a bigger impact than they could individually.

Incorporating environmental communication into curricula can also give learners the opportunity to engage in cooperative learning experiences through experiential learning. Learners can apply what they have learned in real-world settings and get a first-hand understanding of the importance of environmental protection by working together on environmental projects, going on field trips, and taking advantage of other experiential learning opportunities. These experiences not only can help learners retain the material they are learning in the classroom, but they can also give them practical skills like cooperation, problem-solving, and communication—all of which are essential for effective teamwork [5].

In addition, integrating environmental communication into the curriculum can motivate learners to work with local organisations and stakeholders to address environmental issues as well as get involved in their communities. By engaging with experts in the field and participating in community service projects, learners can see the tangible results of their efforts and gain a deeper understanding of the crucial role that cooperation plays in enacting meaningful change [7]. This hands-on approach to environmental communication can give learners a sense of agency and empowerment, which can inspire them to keep working towards a more sustainable future.

7.7.10 Cultivating Knowledge

One of the objectives of integrating environmental communication into curricula for educational institutions is to increase knowledge of pressing environmental issues [8]. Through class discussions, assigned readings, and group projects, learners can learn topics such as pollution, land degradation, climate change, and endangered species. Learners with this knowledge are better equipped to understand the issues facing the environment and the ways that human activity impacts ecosystems. By getting involved in these issues, learners can develop a sense of responsibility and a willingness to take action to protect the environment.

Integrating environmental communication into the curriculum can foster a sense of being part of the natural Africa. Enhancing learners' comprehension of ecosystems, wildlife, and the interconnectedness of all living organisms can facilitate a deeper appreciation of nature's intricacy and splendour. Learners' passion for the natural world might encourage them to advocate for conservation and preservation of the environment. Educators who instil in their learners a sense of awe and wonder for the environment can inspire learners to become stewards of the Earth and take action to protect and preserve our natural resources [11].

Besides, integrating environmental communication into curricula can equip learners with the information and abilities to choose how their actions will impact the environment. By studying sustainable practices, renewable energy sources, and waste reduction strategies, learners can gain the skills and knowledge required to make environmentally conscious decisions in their daily lives [7]. Learners who have a sense of empowerment have the potential to gradually alter their behaviour by developing environmentally friendly and sustainable habits. By giving learners the tools to make a positive environmental impact, educators can create a generation of environmentally conscious individuals who will commit their lives to protecting the environment.

7.8 Case Studies of Initiatives for the Integration of Environmental Issues in Education

There have been many successful initiatives in Africa that have focused on integrating environmental issues into education. Integrating environmental education into the curriculum is essential for promoting sustainable development in Africa.

7.8.1 The Eco-School Programme

Integrating environmental issues into the curriculum is becoming essential as we strive to create a more sustainable future for our world. The Eco-School Programme is an initiative in South Africa aimed at improving environmental awareness and education [15].

The Eco-School Programme is an international initiative aimed at empowering children to engage in environmental conservation. The Wildlife and Environment Society of South Africa (WESSA) oversees the programme in South Africa, successfully promoting environmental education in schools throughout the country [15].

A core component of the Eco-School Programme is the Eco-Schools curriculum, designed to integrate environmental subjects across many disciplines, including science, geography, and social studies. Incorporating environmental subjects into the curriculum enhances students' understanding of the importance of sustainability and conservation.

The Eco-School Programme supplements the curriculum by encouraging practical initiatives among students to improve their school environment. This may include initiatives such as waste management, energy conservation, and the creation of school gardens. By actively engaging in these programmes, children may see the tangible outcomes of their actions and cultivate a sense of environmental responsibility.

Furthermore, the Eco-School Programme provides educators with professional development opportunities to enhance their proficiency in environmental education. This ensures that instructors have the necessary tools to effectively include environmental subjects into their curriculum and foster meaningful discussions on sustainability among students.

7.8.2 Green Clubs

Kenya has seen a rapid increase in environmental degradation due to deforestation, pollution, and other human activities. In response to growing concerns, the government, in partnership with various organisations and educational institutions, has established Green Clubs to promote awareness of environmental issues and advocate for sustainable practices among students [16]. These clubs are crucial for integrating environmental subjects into the curriculum and cultivating a sense of responsibility towards the environment among youth.

The fundamental objective of Green Clubs is to teach young people about the importance of environmental conservation and the impact of human activities on the planet. Students may see the immediate effects of their actions on the environment via activities such as tree planting, waste management programmes, and environmental clean-up projects. Participating in these activities fosters a sense of ownership and accountability for the environment among students, therefore motivating them to implement positive changes in their daily lives.

Furthermore, Green Clubs provide a platform for students to engage in discussions and debates on environmental issues at both local and global levels [16]. This allows students to explore other perspectives, share their ideas, and acquire knowledge from one other's experiences. Green Clubs cultivate critical thinking and problem-solving skills, allowing students to get a deeper understanding of environmental issues and their complexities.

In addition, Green Clubs are vital for integrating environmental education into the academic curriculum. Incorporating topics such as climate change, biodiversity conservation, and sustainable development throughout several curricula allows students to have a thorough understanding of environmental issues. This enhances their understanding of environmental issues and equips them with the essential skills and tools to address these challenges in the future.

7.8.3 Outdoor Education

Outdoor Education in Tanzania is an initiative that aims to provide learners a unique and significant educational experience that combines classroom learning with outside activities in a natural environment [17]. This pedagogical method seeks to promote human growth, cooperation, environmental awareness, and physical health. This also affords learners the opportunity to develop vital life skills, such as communication, problem-solving, leadership, and resilience.

The fundamental aim of outdoor education is to foster a deep connection between learners and the natural environment [17]. Engaging in activities such as hiking, camping, wildlife observation, and conservation efforts allows learners to develop a profound regard for the environment and understand the need of its preservation for future generations. This experiential learning approach cultivates learners' sense of stewardship and environmental responsibility.

Outdoor education is designed to improve physical fitness and overall well-being. Engaging in outdoor activities such as hiking, rock climbing, and canoeing provides learners with the chance to remain active and maintain a healthy lifestyle. These activities aid learners in fostering confidence, improving coordination, and building physical strength, all of which contribute to their general well-being.

In addition to promoting environmental awareness and physical health, Outdoor Education aims to cultivate vital life skills in learners. Engagement in activities like teambuilding exercises, leadership challenges, and problem-solving projects allows learners s to develop interpersonal skills, refine communication talents, and augment their capacity for effective collaboration in group environments. These skills are relevant across all areas of life, including school, career, and interpersonal relationships, and may enhance learners' success in multiple aspects of their lives.

7.8.4 *The Wildlife and Environment Society of South Africa*

The Wildlife and Environment Society of South Africa (WESSA) is a non-profit entity dedicated to promoting environmental conservation and sustainable living practices [29]. One of their main aims is to partner with educational institutions to include environmental issues into the curriculum.

Incorporating environmental education into the curriculum allows students to get a profound understanding of the natural world and the impact of human activities on the environment. This fosters a sense of environmental responsibility and encourages students to make ecologically conscious decisions in their everyday lives.

WESSA aims to provide educators with the resources and support required to integrate environmental issues into their teaching [29]. This includes the delivery of lesson concepts, instructional materials, and professional development opportunities for educators to enhance their proficiency in environmental education. By collaborating with educators, WESSA ensures the effective integration of environmental issues into the curriculum, offering pupils a comprehensive education on these vital topics.

Besides, WESSA is also designed to engage students in environmental education initiatives and projects, while also partnering with educators. This includes the coordination of field trips, seminars, and outreach programmes that facilitate student engagement with environmental issues via experiential learning. By engaging children in these activities, WESSA fosters a passion for environmental conservation and empower adolescents to become environmental stewards in their communities.

Integrating environmental issues into the curriculum improves ecological health and academic achievement for students. Studies demonstrate that students receiving environmental education often achieve higher proficiency in science and have increased environmental knowledge [3]. Incorporating environmental issues into the curriculum allows schools to provide a more holistic and engaging educational experience for students, beyond traditional subjects.

7.9 Conclusion

Environmental communication plays a crucial role in the education system by raising public awareness of environmental issues. By effectively communicating with the public, health communication initiatives can highlight pressing environmental problems and inspire action. Incorporating environmental concepts into school lessons can expose learners to sustainability and conservation from a young age, instilling a sense of responsibility towards the environment. Discussing current environmental challenges can encourage learners to explore innovative solutions and become advocates for change. This hands-on approach can enhance academic success and empower learners to positively impact their communities. Collaboration with stakeholders like community leaders and environmental communication prorgammes can amplify the message, create a culture of environmental stewardship, and promote sustainable practices. Prioritising environmental communication in education, as can be seen in successful initiatives integrating environmental communication in education, is essential for a healthy and sustainable global environment.

Recap

The importance of environmental education in today's society cannot be overemphasised. With the looming threats of pollution, deforestation, and climate change, it is crucial for citizens to understand their responsibility to educate both the current and future generations about the significance of protecting the environment. Knowledge is power, and by equipping individuals with an understanding of environmental issues, environmental communication programmes can motivate them to take meaningful action. It is the collective responsibility of all citizens to promote environmental awareness and encourage sustainable practices through education, ultimately securing a healthier planet for years to come.

Questions

- What is the function of environmental communication in education?
- What is the role that environmental communication in raising public awareness of environmental issues?
- How can environmental communication shape academic curricula?
- What are some of the successful initiatives in integrating environmental communication in education?

References

1. Jacobson MZ. On the causal link between carbon dioxide and air pollution mortality. Geophys Res Lett. 2008;35(3):S2CID 16440166. https://doi.org/10.1029/2007GL031101.
2. World Health Organization. Regional burden of disease due to indoor air pollution; 2013. https://web.archive.org/web/20040701221414/http:/www.who.int/indoorair/health_impacts/burden_regional/en/
3. Cukrowska EM, Yalala BN, Tutu H, Chimuka L. Spatial distribution and source identification of mercury in dust affected by gold mining in Johannesburg, South Africa. In: Proceedings of the international conference of recent trends in environmental science and engineering; 2017. p. 104.
4. Lusilao-Makiese JG, Cukrowska EM, Tessier E, Amouroux D, Weiersbye I. The impact of post gold mining on mercury pollution in the West Rand region, Gauteng, South Africa. J Geochem Explor. 2013;134:111–9.
5. Papu-Zamxaka V, Mathee A, Harpham T, et al. Elevated mercury exposure in communities living alongside the Inanda Dam, South Africa. J Environ Monit. 2010;12(2):472–7.
6. Liousse C, Assamoi E, Criqui P, Granier C, Rosset R. Explosive growth in African combustion emissions from 2005 to 2030. Environ Res Lett. 2014;9(3):035003.
7. Guide on ambient air quality legislation – air pollution series. United Nations Environment Programme; 2023. https://doi.org/10.59117/20.500.11822/42536. ISBN 978-92-807-4029-5.
8. Scorgie Y. Air quality and regulation. National Association for Clean Air; 2013.
9. Kang H, Zhu B, Zhu C, de Leeuw G, Hou X, Gao J. Natural and anthropogenic contributions to long-term variations of SO2, NO2, CO, and AOD over East China. Atmos Res. 2019;215:284–93. https://doi.org/10.1016/j.atmosres.2018.09.012. ISSN 0169-8095
10. Hillstrom K, Hillstrom LC. The worlds environments. A continental overview of environmental issues. Bloomsbury Publishing; 2003. p. 207–8.
11. Fisher S, Bellinger DC, Cropper ML, et al. Air pollution and development in Africa: impacts on health, the economy, and human capital. Lancet Planet Health. 2021;5(10):e681–8. https://doi.org/10.1016/S2542-5196(21)00201-1.
12. Death C. Environmental movements, climate change, and consumption in South Africa. J South Afr Stud. 2014;40(6):1215–34. https://doi.org/10.1080/03057070.2014.964494.
13. Dlamini JS. Safari nation: a social history of the Kruger National Park. Ohio University Press; 2020.
14. Evans HC. Re-articulating media re/presentations of climate change discourse (s) in South Africa: climate change politics in the Global South. Doctoral dissertation. University of KwaZulu-Natal, Howard College; 2019.
15. Evans HC, Dyll L, Teer-Tomaselli R. Communicating climate change: theories and perspectives. Handbook of climate change communication: Vol. 1: theory of climate change communication. Springer International Publishing; 2018. p. 107–22.
16. Leal Filho W, Manolas E, Azul AM, Azeiteiro UM, McGhie H. Handbook of climate change communication: Vol. 2. Practice of climate change communication. Springer; 2018.
17. Francis D, Valodia I, Webster E. Inequality studies from the global south. London: Routledge; 2020. https://doi.org/10.4324/9780429282447.

18. Francis D, Webster E. Poverty and inequality in South Africa: critical reflections. Dev South Afr. 2019;36(6):788–802. https://doi.org/10.1080/0376835X.2019.1666703.
19. Ghosh A. Can the non-human speak? Other beings in myth, literature and ethnography. Yale NUS College; 2019. https://www.yale-nus.edu.sg/events/amitav-ghosh-can-the-non-human-speak/
20. Ghosh A. The great derangement: climate change and the unthinkable. UK: Penguin; 2018. https://www.bibliovault.org/BV.landing.epl?ISBN=9780226323176
21. Grebowicz M. Glacial time and lonely crowds: the social effects of climate change as internet spectacle. EH. 2014;5(1):1–11. https://doi.org/10.1215/22011919-3615388.
22. Green L. Rock| water| life: ecology and humanities for a Decolonial South Africa. Duke University Press; 2020.
23. Kamau S. The significance of oral history in sustainable development: perspectives from East Africa. J Sust Dev. 2016;4(1):112–25.
24. Ibikunle O. Community proclamations on environmental preservation in Africa. Afr Dev Stud. 2018;33(4):201–15.
25. Dube S. Proclamations and announcements: communication tools for environmental advocacy in Africa. Afr J Comm Stud. 2024;30(2):113–25.
26. Odhiambo B. The role of proclamations in raising awareness about environmental issues in Africa. J Commun Stud. 2017;14(5):67–79.
27. Okonkwo N. The wisdom of African elders in environmental communication. ESR. 2020;18(3):112–25.
28. Chigunta P. Tribal meetings: a platform for environmental dialogue in African communities. J Environ Comm. 2015;20(4):389–402.
29. Francis D, Webster E. Poverty and inequality in South Africa: critical reflections. Dev South Afr. 2019;36(6):788–802. https://doi.org/10.1080/0376835X.2019.166670.

CHAPTER 8

Monitoring and Evaluating the Impact of Environmental Communication Strategies

Introduction
Environmental communication programmes play a crucial role in raising awareness about various environmental issues and encouraging individuals to adopt sustainable behaviours. However, simply launching these campaigns is not enough to ensure their effectiveness. Monitoring and evaluation of these campaigns is equally crucial to determine their impact and make necessary adjustments to attain better results. This chapter deals with the monitoring and evaluation of the impact of environmental communication strategies. This chapter starts by covering the importance of monitoring and evaluation of environmental communication programmes. Then the chapter discusses the theoretical metrics for monitoring and evaluating environmental communication programmes. The chapter then discusses some conceptual metrics that can be used to monitor and evaluate environmental communication programmes.

8.1 The Importance of Monitoring and Evaluating Environmental Communication Programmes

Monitoring and evaluation of environmental communication programmes is essential to ensure their effectiveness and maximise their impact [1]. By careful monitoring of key indicators, measuring changes in behaviour, and uncovering unexpected results, environmental communication programmes can make informed decisions about their communication and strategy approach. Ultimately, monitoring and evaluation are essential tools for advancing environmental sustainability and achieving long-lasting change in communities.

© The Author(s), under exclusive license to Springer Nature Switzerland AG 2025

G. Mutinta, *An African Perspective on Environmental Communication*, https://doi.org/10.1007/978-3-032-05052-6_8

8.1.1 Impact on the Target Audience

Monitoring and evaluation are essential to find out how environmental conservation communication programmes are impacting the target audience in Africa [1]. For example, monitoring and evaluating environmental communication programmes provides valuable insights into the overall effectiveness of the programmes. Programmes assessors can evaluate key performance indicators such as changes in environmental conservation habits to determine if programmes are successfully achieving their goals. With the use of this data, stakeholders can identify the programmes' strong and weak aspects and determine what has to be altered or improved.

Environmental communication programmes can be monitored and evaluated to see how well they engage and reach the intended audience. Programmes assessors can evaluate indicators such as participant demographics, programmes satisfaction, and attendance rates to determine whether the programmes are reaching the most vulnerable to environmental problems. This information is necessary to ensure that resources are allocated efficiently and that the programme is tailored to the needs of the intended audience.

Evaluation of sustainability and long-term outcomes is also made feasible by monitoring and evaluating environmental communication programmes. Assessors can determine if programmes are contributing to participants' long-term behaviour and well-being by keeping an eye on outcomes such as maintenance of environments, modifications to lifestyle choices, and reductions in environmental-related issues. This information is essential for directing the development of new programmes and ensuring that funding is given to projects that have a major and enduring impact.

In addition, monitoring and evaluating environmental communication programmes demonstrate environmental organisations' transparency and accountability to stakeholders. Programmes evaluators may build credibility and trust by providing factual information on programmes outcomes, impact, and cost-effectiveness to funders, lawmakers, and the community. This transparency is essential to sustaining programmes support, securing further funding, and ultimately triumphing in the protracted war against environmental destruction.

8.1.2 Identifying Potential Barriers

Implementing environmental interventions alone is not enough. It is crucial to monitor and evaluate their effectiveness in order to identify potential roadblocks to success [2]. By providing insights into the programme's involvement and reach, monitoring and evaluating environmental communication programmes can help uncover potential obstacles to success. By analysing data on programme participation and engagement levels, it is possible to determine whether particular populations are being underserved or are not connecting with the programme as intended. With this knowledge, programme managers can better tailor their interventions to the needs of these populations and increase the programme's overall efficacy.

Additionally, monitoring and evaluating environmental communication efforts can help identify any barriers to the implementation and delivery of the programme. By getting feedback from programme implementers and participants, it is possible to discover a lot about the programme's advantages and possible areas for improvement. For example, if participants regularly report difficulty accessing programme materials or find some programme components confusing or unhelpful, programme managers can make the necessary adjustments to remove these barriers and increase programme effectiveness.

Furthermore, monitoring and evaluating environmental communication programmes efforts may help identify any obstacles with regard to the outcomes and impact of the campaigns. It is possible to assess the programme's effectiveness in promoting behaviour modification and attaining the targeted environmental goals by keeping an eye on key performance indicators such alterations environmental behaviour, and conversation patterns. Programme managers can use this data to pinpoint areas that need improvement and adjust their approach if results do not meet their expectations.

Monitoring and evaluating environmental communication programmes can help identify potential roadblocks to sustainability and scalability. By assessing the programme's long-term impacts and feasibility for wider adoption, it is possible to determine whether it is feasible in the long run and whether it can be expanded to serve a larger population. Programme managers should make an effort to address any challenges related to scalability or sustainability to ensure that the programme can continue to have a lasting impact on the environment.

8.1.3 Maximising Resources and Cost-Effectiveness

Monitoring and evaluating environmental communication programmes is essential to maximising resources and cost-effectiveness [3–5]. This includes identifying the programme's strong and weak aspects in terms of execution and design. Through ongoing monitoring and evaluation, programme managers may determine what aspects of the programme are working effectively and what needs improvement. By identifying strengths, resources can be allocated more effectively to build on the programme's successful components. Like strengths, weaknesses can also be worked upon, leading to a more efficient and cost-effective programme in the end.

Additionally, monitoring and evaluating environmental communication programmes can help ensure that resources are distributed as effectively and efficiently as feasible. Using programme impact analysis and outcome monitoring, managers can identify areas where resources may be misapplied or handled improperly. Resource allocation may therefore be maximised with the use of this data, ensuring that funds are allocated to the strategies and initiatives that have the highest possibility of reducing environmental degradation rates. This targeted approach can maximise the programmes' impact while cutting down on resource waste.

Besides, furthermore, monitoring and evaluating environmental communication programmes efforts can help with future programme design and decision-making. By compiling information on programme outcomes and effects, programme managers can make evidence-based decisions on the design and implementation of their programmes. Programme approaches can be quickly adjusted based on what is and is not functioning using this data. By frequently evaluating the programme, administrators may make necessary modifications and improvements to make it more responsive to the needs of the target audience. Eventually, this will lead to more cost-effective and successful results.

Monitoring and evaluating environmental communication programmes can strengthen the case for continued funding and support. Programme administrators may make a compelling argument for continued funding of environmental communication programmes by supplying information and outcomes that demonstrate the impact and effectiveness of the programme. This can help secure funding and support from stakeholders and decision-makers, which can help assure the programme's longevity. Administrators can make a strong case for the funding of environmental communication programmes by highlighting the advantages and cost-effectiveness of the endeavour.

8.1.4 Building a Strong Evidence Base

Monitoring and evaluating environmental communication programmes efforts can help provide a strong evidence base for future interventions [6–8]. For example, monitoring and evaluating environmental communication programmes efforts can allow people to assess the effectiveness of different environmental communication initiatives. By tracking key indicators, it is possible to determine which policies are most successful in promoting conservative behaviours and preventing environmental degradation. This information may then be used to steer the creation of later interventions, ensuring that financial resources are allocated to projects that have proven successful.

By monitoring and evaluating environmental communication programmes efforts, it is possible to identify areas that require improvement. Environmental communication organisations can find any weaknesses or inadequacies in the current environmental communication approach by looking at the outcomes of these initiatives. The previously described information can then be utilised to apply targeted adjustments to ongoing projects or to develop new strategies that address the identified shortcomings. Therefore, monitoring and evaluating environmental communication programmes efforts can help to advance the fight against environmental degradation over time.

It is possible to determine what makes environmental communication strategies successful or unproductive by monitoring and evaluating these efforts. By keeping an eye on a number of variables, such as social support levels, the availability of environmentally friendly options, and socioeconomic status, people can gain a deeper understanding of the elements that lead to environmental problems and the barriers that prevent behaviour change. With this knowledge, interventions can be more successfully customised to the particular needs and circumstances of different groups, increasing the likelihood that they will be successful.

Finally, monitoring and evaluating environmental communication programmes efforts can help organisations demonstrate their value to key stakeholders including lawmakers, funders, and local citizens. By presenting strong evidence of the effectiveness of environmental communication initiatives, it is possible to secure the money needed to continue and grow existing efforts and boost support for upcoming initiatives. By building a strong evidence base for environmental communication interventions, environmental organisations can ultimately ensure that money is directed towards projects that have the highest possibility of considerably enhancing the environment and welfare of individuals and communities.

8.1.5 *Ascertaining the Impact of the Messages and Communication Strategies*

One of the key benefits of monitoring and evaluating environmental communication programmes is the ability to measure the impact of messages on target audiences [9–11]. By tracking indicators such as message reach, audience engagement, and behaviour change, organisations can gauge the effectiveness of their communication efforts. This information can help organisations identify which messages are resonating with audiences and which are not, allowing for adjustments to be made to improve overall impact.

Furthermore, monitoring and evaluating environmental communication programmes can help organisations assess the success of their communication strategies. By analysing data on message dissemination, audience feedback, and campaign outcomes, organisations can determine what is working well and what can be improved upon. This information can be used to refine communication strategies, ensuring that messages are delivered in a way that is most effective in reaching and engaging target audiences.

In addition, monitoring and evaluating environmental communication programmes can help organisations demonstrate the value of their communication efforts to stakeholders. By providing evidence of the impact of communication programmes, organisations can show that their efforts are making a difference in raising awareness and promoting action on environmental issues. This can help secure continued support and funding for communication initiatives, ensuring that they can continue to make a positive impact in the future.

8.1.6 *Promoting Accountability and Transparency*

Monitoring and evaluating environmental communication programmes may encourage transparency and accountability by providing stakeholders with actual data and evidence of the programme's success [12]. By collecting information on key indicators, organisations can ascertain whether a programme is actually benefiting the lives of its participants. Transparency in data may help stakeholders understand the true impact of the programme and hold programme implementers responsible for the results.

Research also shows that monitoring and evaluation can help identify chances for programming improvement. Through data collection throughout the monitoring and assessment phase, organisations can determine which programme aspects are not yielding the desired results. With this information, the programme may then be enhanced and modified as necessary, ensuring that resources are used sensibly and efficiently. This continuous improvement strategy cannot only increase the programme's performance but also demonstrate to stakeholders the organisation's commitment to transparency and accountability.

Furthermore, monitoring and evaluating environmental communication programmes can help ensure that money is spent wisely. By comparing the programme's outcomes to the resources spent, organisations may determine if a programme is economical and whether resources are being used properly. This level of transparency in resource allocation can help to build confidence among stakeholders and establish a commitment to responsibility in the management of public finances.

Finally, monitoring and evaluating environmental communication programmes efforts might help foster a culture of innovation and learning. Organisations can examine the data collected throughout the monitoring and evaluation process to learn from both their successes and their failures. This can assist them in the future in making better judgements and implementing strategies that are supported by evidence. This commitment to learning and innovation not only increases programme performance but also demonstrates the openness of decision-making processes.

8.1.7 *Opportunities of Innovation*

One of the key benefits of monitoring and evaluating environmental communication programmes is that it can lead to the identification of opportunities for innovation [13–15]. By analysing data on audience engagement, message effectiveness, and programme impact, organisations can gain valuable insights into what resonates with their target audience and what does not. This information can then be used to develop new and creative ways of communicating environmental messages that are more engaging and impactful.

For example, if monitoring and evaluation reveal that a particular communication strategy is not effectively reaching a key demographic, organisations can use this information to experiment with different approaches or channels to better reach that audience. This could involve partnering with influencers, leveraging social media platforms, or using interactive technologies to make the message more accessible and engaging.

Innovation is key to keeping environmental communication programmes relevant and impactful in a rapidly changing world. By regularly monitoring and evaluating these programmes, organisations can stay ahead of the curve and continuously adapt to new trends and technologies. This not only helps to ensure that the message remains fresh and relevant but also maximises the potential for creating real change in behaviours and attitudes towards the environment.

8.1.8 Sustaining Long-Term Impacts

To ensure that environmental communication programmes have the intended long-term benefits on people's lives and communities, they must be continuously monitored and assessed [16]. Monitoring and evaluating environmental communication activities allows one to track progress and identify areas of weakness in treatment implementation. Environmental organisations can ensure that resources are being used effectively and efficiently to achieve the desired impacts by collecting and analysing data on programme activities, outcomes, and outputs consistently. Programme managers can utilise this data to influence resource allocation choices and to make the necessary adjustments to improve programme delivery.

Environmental communication programmes are monitored and evaluated to demonstrate their impact on individuals and communities. Environmental organisations may monitor changes in behaviour, attitudes, and environmental outcomes to determine whether the programme is achieving its intended goals and objectives. These changes can demonstrate the advantages of supporting programmes aimed at preventing environmental degradation and boost support for the effort among lawmakers, donors, and people of the community.

Besides, environmental communication initiatives may also draw lessons from both their successes and failures. By identifying what is and is not working effectively, they can make the required changes and enhancements to boost the efficacy of programme activities. This continuous learning process is necessary to maintain long-term impacts as well as the relevance and responsiveness of environmental communication initiatives to the changing requirements of individuals and communities.

By giving regular updates on the status and accomplishments of environmental communication programmes, environmental organisations can demonstrate their dedication to achieving results and accepting accountability for our actions. With time, this transparency can help maintain support for the initiative by building confidence among participants.

Monitoring and evaluating environmental communication programmes present an opportunity for collaboration and knowledge-sharing. By combining our resources, expertise, and best practices, as well as exchanging information and best practices with other organisations and initiatives, people can tackle the complex and multifaceted issue of environmental degradation. In the end, by building a more solid and well-coordinated response to environmental communication measures, this cooperation may have more profound long-term benefits on individuals and communities.

8.2 Theoretical Monitoring and Evaluation

Environmental communication programmes are essential for raising awareness, promoting change, and ultimately protecting our planet. However, simply implementing these programmes is not enough—there is need to also monitor and evaluate their effectiveness to ensure they are achieving their intended goals [17, 18]. There are several theories that can be used to assess the impact of environmental communication programmes, and it is important for organisations to utilise these tools to make informed decisions and improve their efforts.

8.2.1 The Health Belief Model

One of the frameworks that can be utilised in monitoring and evaluating communication programmes that are raising awareness and promoting behaviour change towards sustainability and environmental protection initiatives is the health belief model. The health belief model is a widely used theoretical framework in the field of health communication, which can be adapted and applied to the context of environmental communication [19].

One of the main components of the health belief model is perceived susceptibility, which refers to an individual's belief about the likelihood of being affected by a health condition. In the context of environmental communication initiatives, perceived susceptibility can be translated to how individuals perceive the risks and threats posed by environmental issues such as climate change, pollution, and deforestation. By assessing and monitoring changes in individuals' perceived susceptibility through surveys, interviews, or focus groups, organisations can determine the effectiveness of their communication strategies in conveying the urgency and importance of environmental protection (Fig. 8.1).

Another element of the health belief model is perceived severity, which refers to an individual's belief about the seriousness of a health condition [19]. In the context of environmental communication initiatives, perceived severity can be translated to how individuals perceive the potential consequences of failing to take action towards environmental sustainability. By evaluating changes in individuals' perceived severity through pre- and post-campaign surveys or content analysis of communication materials, organisations can assess whether their messaging is effectively conveying the seriousness of environmental issues and motivating individuals to take action.

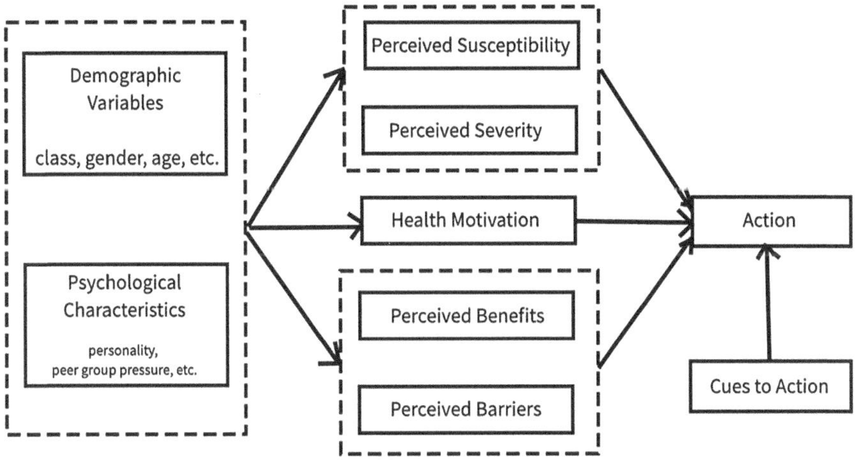

Fig. 8.1 The health belief model. Author (2025)

Furthermore, the health belief model emphasises the role of perceived benefits and barriers in influencing individuals' behaviour [19]. In the context of environmental communication initiatives, perceived benefits can refer to the rewards and positive outcomes of engaging in sustainable behaviours, while perceived barriers can refer to the obstacles and challenges that individuals may face in adopting these behaviours. By monitoring changes in individuals' perceived benefits and barriers through qualitative research methods or surveys, organisations can identify gaps in their communication strategies and address them to better motivate behaviour change towards sustainability.

In addition to these components, the health belief model also highlights the importance of cues to action, which are triggers that prompt individuals to take action towards protecting their health [19]. In the context of environmental communication initiatives, cues to action can include calls to action in communication materials, reminders about the importance of sustainable behaviours, or prompting individuals to take specific actions such as recycling or reducing energy consumption. By monitoring the effectiveness of cues to action through tracking website engagement, social media metrics, or campaign reach, organisations can assess the impact of their communication efforts in prompting behaviour change towards sustainability.

8.2.2 The Stages of Change Model

In order to effectively evaluate and monitor the success of these initiatives, the stages of change model can be a useful tool. The stages of change model, also known as the transtheoretical model, provide a framework for understanding how individuals progress through different stages of behaviour change [20].

The first stage of the model is pre-contemplation, where individuals are not yet considering changing their behaviour. In the context of environmental communication initiatives, this stage could represent individuals who are unaware of or indifferent towards environmental issues. Monitoring and evaluation efforts at this stage could involve measuring baseline levels of awareness and attitudes towards environmental conservation (Fig. 8.2).

The second stage is contemplation, where individuals start to consider changing their behaviour [20]. This could be reflected in environmental communication initiatives by assessing levels of interest and engagement with environmental issues. Surveys, focus groups, and social media analytics can be used to track how many individuals are actively thinking about making changes to their behaviour in relation to environmental sustainability.

The third stage is preparation, where individuals are ready to take action towards behaviour change [20]. In the context of environmental communication initiatives, this could involve tracking the number of individuals who are actively seeking out information and resources to support their efforts towards environmental conservation. Monitoring and evaluation efforts at this stage could also involve assessing the effectiveness of different communication channels and messaging strategies in encouraging action.

The fourth stage is action, where individuals start to implement changes to their behaviour [20]. This stage could be evaluated by tracking participation rates in environmental initiatives, such as recycling programmes, community clean-up events, or sustainable living workshops. Monitoring the impact of these actions on the environment, such as reductions in waste or carbon emissions, can also provide valuable data for evaluating the success of environmental communication initiatives.

The final stage is maintenance, where individuals work to sustain their behaviour changes over time [20]. Monitoring and evaluation efforts in this

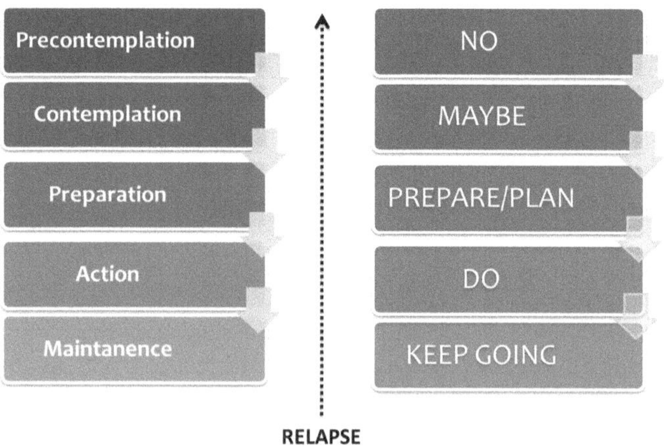

Fig. 8.2 Stages of change model. Author (2025)

stage could involve tracking long-term behaviour change and measuring the ongoing impact of environmental communication initiatives on individuals and communities. Surveys, interviews, and case studies can be used to assess the lasting effects of these initiatives on attitudes, beliefs, and behaviours related to environmental sustainability.

8.2.3 *Knowledge-Deficit Model*

The knowledge-deficit model is a common approach used to explain how individuals process information and make decisions based on the lack of knowledge or understanding of a particular subject [21]. When it comes to monitoring and evaluating environmental communication initiatives, the knowledge-deficit model can be a valuable tool in assessing the effectiveness of these initiatives in educating and engaging the public on environmental issues.

One way the knowledge-deficit model can be utilised in monitoring environmental communication initiatives is by measuring the level of knowledge and understanding the target audience before and after the initiative. By conducting surveys or assessments, organisations can determine if their messages are effectively increasing awareness and knowledge among the public. If there is a significant increase in knowledge after the initiative, it suggests that the communication strategies were successful in educating and engaging the audience.

In addition, the knowledge-deficit model can also be used to evaluate the impact of environmental communication initiatives on changing attitudes and behaviours towards the environment. By examining whether there is a shift in attitudes or behaviours before and after the initiative, organisations can determine if their messages are influencing the target audience to take action. For example, if a campaign to reduce plastic waste results in a decrease in the use of single-use plastics, it indicates that the communication strategies were effective in triggering behaviour change.

Furthermore, the knowledge-deficit model can be used to identify gaps in knowledge and understanding that may exist within the target audience. By analysing the results of surveys or assessments, organisations can tailor their communication strategies to address specific knowledge deficits and enhance the effectiveness of their initiatives. For example, if a survey reveals that the public has a limited understanding of the impact of climate change on biodiversity, organisations can focus their communication efforts on providing more information on this topic.

8.2.4 *Social Marketing Model*

The social marketing model is a strategic approach that utilises marketing principles to promote social causes and behaviour change [22]. By applying this model to environmental communication initiatives, organisations can better understand their target audience, identify key messaging strategies, and measure the impact of their efforts.

One key component of the social marketing model is audience segmentation [23]. By dividing the target audience into specific groups based on demographics, attitudes, and behaviours, organisations can tailor their messaging to effectively reach each segment. For example, a campaign aimed at reducing plastic pollution may target environmentally conscious consumers who are already receptive to the message, while also reaching out to more sceptical individuals with information on the environmental impact of plastic waste.

Furthermore, the social marketing model emphasises the importance of clear communication objectives and key messages [24]. By establishing clear goals for the campaign and developing compelling messaging that resonates with the target audience, organisations can increase the likelihood of success. For instance, a campaign promoting sustainable living practices may focus on the benefits of reducing carbon emissions and saving money on energy bills, in order to appeal to both eco-conscious consumers and cost-conscious individuals.

In addition, the social marketing model emphasises the use of feedback and evaluation to measure the impact of environmental communication initiatives [24]. By collecting data on key performance indicators such as reach, engagement, and behaviour change, organisations can assess the effectiveness of their campaigns and make informed decisions on future strategies. For example, a campaign promoting recycling could track the number of participants who sign up for a recycling programme, as well as the amount of waste diverted from landfills as a result of the campaign.

8.2.5 The Diffusion of Innovation Theory

The diffusion of innovation theory is a powerful tool that can be used in monitoring and evaluating environmental communication initiatives. By understanding how new ideas, products, or practices spread through a population, organisations can better tailor their messaging and strategies to effectively engage with target audiences and drive positive environmental outcomes.

One key aspect of the diffusion of innovation theory is identifying different groups within a population based on their willingness and ability to adopt new ideas or behaviours [25]. This can help environmental communication initiatives to target their messaging towards early adopters and influencers who can help spread the word to a wider audience. By monitoring the uptake of new ideas within these groups, organisations can track the progress of their initiatives and make adjustments as needed to increase adoption rates (Fig. 8.3).

Another important aspect of the diffusion of innovation theory is the notion of social norms and peer influence. People are more likely to adopt new behaviours if they see others around them doing the same. By leveraging social networks and peer-to-peer communication, environmental communication

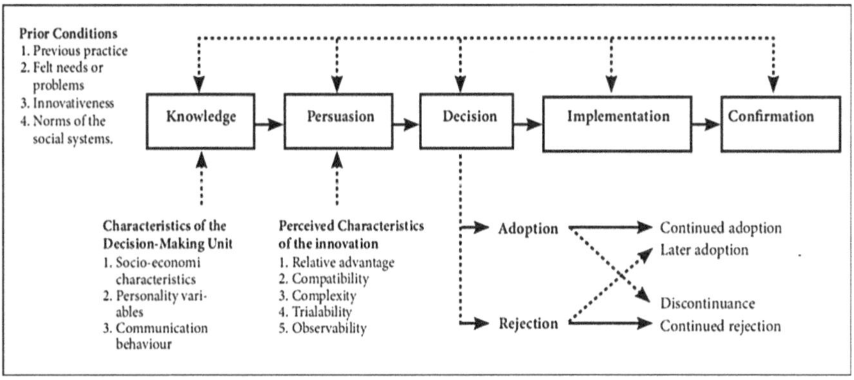

Fig. 8.3 The diffusion of innovation theory. Author (2025)

initiatives can create a ripple effect that amplifies their impact and accelerates the adoption of sustainable behaviours.

Monitoring and evaluating environmental communication initiatives using the diffusion of innovation theory can provide valuable insights into the effectiveness of different communication channels, messages, and strategies. By tracking the spread of new ideas and behaviours through a population, organisations can identify which approaches are most successful and which may need to be adjusted or discontinued [26].

8.2.6 The Framing Theory

The framing theory is a powerful tool that can be utilised in monitoring and evaluating environmental communication initiatives [27]. By understanding how messages are framed and presented to the public, we can better gauge the effectiveness of these initiatives in promoting environmental awareness and action.

One main aspect of the framing theory is understanding how different frames can shape public perceptions of environmental issues [28]. Frames are essentially the lenses through which individuals view an issue, and they can have a significant impact on how the message is received and interpreted. For example, framing an environmental issue as a health concern may resonate more with certain audiences than framing it as an economic issue. By utilising different frames in environmental communication initiatives, we can tailor our messages to different target audiences and increase the chances of reaching and engaging them effectively.

In monitoring and evaluating environmental communication initiatives, the framing theory can help us assess the impact of these initiatives on public opinion and behaviour. By examining how different frames are received by the public and measuring changes in attitudes and behaviours over time, we can determine which frames are most effective in promoting environmental awareness and action. This information can then be used to refine and improve future communication efforts, ensuring that they are as impactful as possible.

Furthermore, the framing theory can also help organisations identify potential barriers to effective communication and engagement [29]. By understanding how certain frames may be perceived negatively or misunderstood by the public, we can address these challenges and adjust our messaging accordingly. This can help us avoid unintended consequences and ensure that our communication initiatives are as successful as possible in promoting positive environmental outcomes.

8.2.7 The Agenda-Setting Theory

The agenda-setting theory is a powerful tool that can be utilised in monitoring and evaluating environmental communication initiatives. This theory posits that the media has the ability to shape the public's perception of issues by deciding which topics are important and which are not [30]. By understanding and implementing this theory, organisations can effectively track the progress of their environmental communication efforts and make data-driven decisions to improve their impact (Fig. 8.4).

One way in which the agenda-setting theory can be used in monitoring and evaluating environmental communication initiatives is by analysing media coverage [31]. By examining the types of environmental issues that are being reported on and the frequency with which they are being discussed, organisations can gain valuable insights into how well their messaging is being received by the public. If certain topics are consistently being ignored or misrepresented in the media, it may be an indication that the organisation needs to adjust their communication strategy to better align with the public's interests and concerns.

Additionally, the agenda-setting theory can help organisations track changes in public opinion over time [32]. By comparing media coverage of environmental issues with public opinion polls, organisations can identify trends and

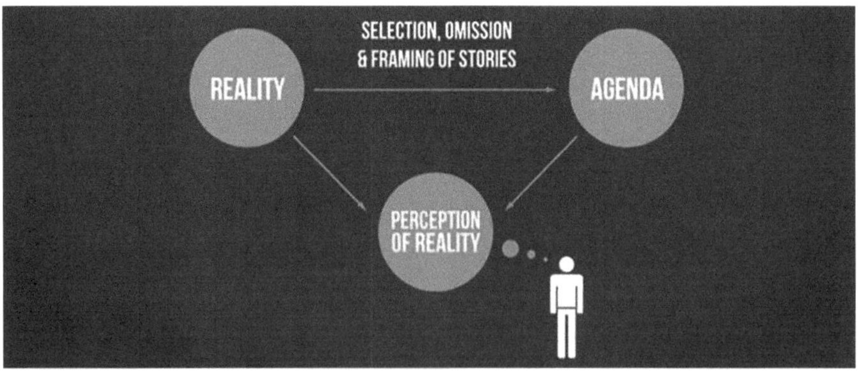

Fig. 8.4 The agenda-setting theory. Author (2025)

patterns that can inform future communication efforts. For example, if media coverage of climate change increases significantly but public concern remains stagnant, it may be a sign that the organisation needs to find new ways to engage and educate the public on this critical issue.

Furthermore, the agenda-setting theory can be used to assess the effectiveness of specific communication tactics and messaging strategies [32]. By conducting surveys or focus groups to measure awareness and understanding of key environmental issues before and after implementing a communication campaign, organisations can gauge the impact of their efforts and make adjustments as needed. For example, if a campaign focused on recycling is launched but fails to increase public participation in recycling programmes, it may be a signal that the messaging needs to be revised to better resonate with the target audience.

8.2.8 *The Participatory Communication Model*

The participatory communication model provides a valuable framework for monitoring and evaluating environmental communication initiatives as it involves engaging all stakeholders in the communication process [33].

One of the key features of the participatory communication model is its emphasis on involving all relevant stakeholders in the communication process [34]. This includes not only the creators and producers of the communication initiatives but also the target audience, community members, and other interested parties. By including the perspectives and feedback of all stakeholders, the participatory communication model allows for a more comprehensive and accurate assessment of the impact of environmental communication initiatives.

When using the participatory communication model to monitor and evaluate environmental communication initiatives, it is important to involve the target audience in the process. This can be done through focus group discussions, surveys, and other participatory methods that allow for direct feedback from the audience. By including the perspectives and opinions of the target audience, organisations can gain valuable insights into the effectiveness of their communication strategies and make necessary adjustments to improve their impact.

In addition to engaging the target audience, the participatory communication model also emphasises the importance of involving community members and other stakeholders in the monitoring and evaluation process [35]. By including the perspectives of community members, organisations can gain a better understanding of the broader impact of their communication initiatives on the local environment and community dynamics. This can help organisations identify potential barriers to behaviour change and develop more effective strategies for addressing them.

Furthermore, the participatory communication model promotes a collaborative approach to monitoring and evaluating environmental communication initiatives [35]. By working together with all stakeholders, organisations can foster a sense of ownership and shared responsibility for the success of the initiatives. This can lead to greater commitment and engagement from all parties involved, ultimately increasing the likelihood of success in achieving the desired outcomes.

8.3 Conceptual Metrics for Assessing the Impact of Environmental Communication Programmes

Evaluating the effectiveness of environmental communication programmes requires measuring their impact using key performance indicators. By looking at these metrics, organisations may discover areas for improvement and get crucial knowledge about the success of their communication campaigns (Fig. 8.5).

8.3.1 Awareness

One metric used to assess the success of environmental communication campaigns is awareness [36]. Programme managers can evaluate their effectiveness in disseminating important environmental information by measuring the degree of awareness among the target group. High awareness levels indicate that the message has been effectively communicated and has reached a sizeable audience, whilst low awareness levels may indicate that the messaging or distribution strategies need to be adjusted.

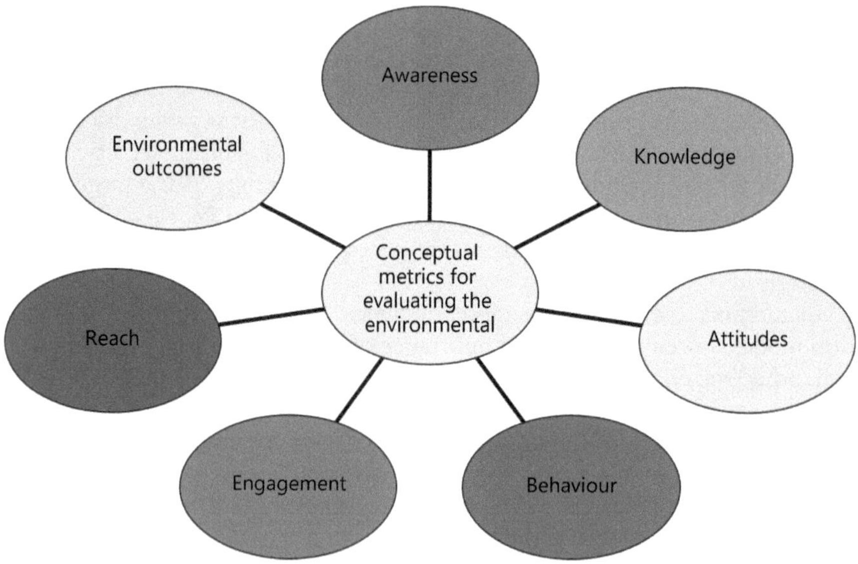

Fig. 8.5 Conceptual metrics. Author (2025)

Awareness can provide insight into how well the intended audience understands and retains environmental information. It is not enough for people to just know about environmental problems or recommended practices; they also need to understand the importance of the information and be able to recall it over time. Campaign managers can compare awareness levels before and after an environmental communication campaign to see if the audience has retained and assimilated the information.

By evaluating awareness, environmental communication programmes may be assessed for their impact on attitudes and behaviours. A better knowledge of environmental issues might lead to a change in attitudes and actions that would enhance environmental outcomes. Programme managers may monitor awareness levels together with changes in attitudes and behaviours to assess how successfully their environmental communication programmes are producing positive environmental-related outcomes.

The whole efficacy and impact of an environmental communication campaign may be substituted with awareness. High levels of awareness indicate that the programme has been effective in capturing the attention and engagement of the target audience, whilst low levels may indicate the need for adjustments or additional resources. By tracking awareness levels over time and comparing them with other important indicators, programme managers may evaluate the overall effectiveness and impact of their environmental communication programmes.

8.3.2 Knowledge

Knowledge is a critical indicator of the target audience's comprehension of the information provided by environmental communication initiatives [36]. By assessing participants' knowledge prior to and during their exposure to the programme, organisations may ascertain how effectively they have absorbed the content being presented. This makes it possible for us to assess how well the programme is doing at disseminating new knowledge and raising public awareness of issues connected to environmental problems.

Knowledge is one way to gauge how environmental communication efforts have affected people's behaviours. Following up on new knowledge on environmental-related topics, people are more likely to change their behaviour. For example, after seeing a presentation that educates them about the risks of cutting trees, people are more inclined to reduce or stop cutting trees. By keeping an eye on changes in behaviour and knowledge, environmental organisations can evaluate the programme's effectiveness in promoting healthier behaviours.

Moreover, knowledge among participants in environmental communication initiatives indicates agency and empowerment. When people are knowledgeable about their environment and well-being, they are better equipped to make decisions about their care and take action to enhance their quality of life. By assessing the knowledge that participants have gained, organisations can determine the degree to which they feel prepared to take control of their environment and make positive life choices.

Information may be utilised to evaluate the overall effectiveness and long-term sustainability of environmental communication programmes. By monitoring how long individuals retain knowledge, organisations can assess if the messages have had a lasting impact on people's understanding and awareness of environmental issues. This makes it possible for us to evaluate how well the programme changed the knowledge and behaviour of the target audience in the long run.

8.3.3 Attitudes

Individuals' attitude towards the environment provides insight into their conceptions of the environmental problems [37]. By comparing attitudes before and after exposure to a programme, environmental communication organisations can assess the degree to which it has impacted people's ideas and perspectives. For example, if a significant portion of the target group had negative attitudes on throwing litter in the bin every day before the beginning of the programme, a positive change in opinions following the campaign implies that the messaging has been successful in changing perceptions.

Attitudes can indicate a change in behaviour [38]. Research has shown that behavioural intentions, behaviours, and attitudes are highly correlated. By recognising how attitudes regarding environmental issues influence behaviour, environmental communication strategies may be developed to effectively target and address significant attitudes that may be barriers to adopting environmentally friendly habits. If a campaign's goal is to promote the use degradable bags, for instance, assessing attitudes about degradables might help identify potential obstacles and adjust the messaging to address them.

Moreover, attitudes provide valuable insights into the overall effectiveness of environmental communication programmes. By monitoring changes in attitudes over time, programme evaluators can ascertain which communications tactics work best and which may need to be revised. This feedback loop allows for ongoing improvement of communication efforts to ensure that they are reaching their intended audience and effecting positive change.

And last, attitudes serve as a barometer for the overall impact that environmental communication efforts have on individuals and groups. It is possible to assess the long-term impacts of communication initiatives on public environmental outcomes by keeping an eye on changes in public perceptions of environmental-related behaviours, regulations, or services. Positive attitude shifts indicate that the programme has been successful in raising awareness, transforming people's viewpoints, and ultimately changing behaviour to enhance environmental outcomes.

8.3.4 Behaviour Change

To ensure the success of environmental communication campaigns, it is important to use behaviour change as a metric to monitor and evaluate their impact [39, 40]. Behaviour change is a powerful indicator of how effective a campaign has been in raising awareness and motivating people to take action. By monitoring changes in behaviour, environmental communication organisations can gain valuable insights into the effectiveness of their communication strategies and make informed decisions on how to improve them.

For example, if a campaign aims to encourage people to reduce their plastic waste, environmental communication organisations can measure the changes in people's behaviour such as using reusable bags or bottles, recycling more frequently, or participating in beach clean-ups. By tracking these changes over time, environmental communication organisations can determine whether the campaign has been successful in influencing people's habits and attitudes towards plastic consumption.

Behaviour change can also help organisations to identify gaps in communication efforts and tailor our messages more effectively. For instance, if a campaign targeting water conservation is not yielding any significant changes in behaviour, environmental communication organisations can reassess our messaging and strategy to better resonate with the target audience.

Moreover, behaviour change can serve as a key performance indicator for environmental communication campaigns, providing tangible evidence of their impact and effectiveness. By setting specific behaviour change goals at the outset of a campaign, we can easily track progress and evaluate the success of our efforts.

8.3.5 Engagement

Engagement is a crucial metric when it comes to monitoring and evaluating environmental communication campaigns [41]. By measuring the level of engagement with the target audience, organisations can gain valuable insights into the effectiveness of their messaging and ultimately improve their impact on environmental issues.

For example, engagement can be used as a metric is through social media interactions. Platforms like Facebook, Twitter, and Instagram provide valuable data on the number of likes, shares, comments, and retweets that a campaign receives. By analysing this data, organisations can determine which messages are resonating with their audience and adjust their communication strategies accordingly.

Another way to measure engagement is through website traffic and online interactions. By tracking metrics such as the number of visitors to a campaign website, the length of time spent on the site, and the pages viewed, organisations can gain a better understanding of how effectively their message is reaching their target audience. Additionally, measuring the number of people who sign up for newsletters or take action on the website (such as signing a petition or making a donation) can provide further insights into the impact of the campaign.

Furthermore, engagement can also be measured through offline interactions, such as attendance at events, participation in community clean-up initiatives, or feedback from focus groups and surveys. By gauging the level of participation and interest in these activities, organisations can assess the effectiveness of their communication strategies and make adjustments as needed.

8.3.6 *Reach*

To measure the effectiveness of environmental communication campaigns, reach can be used as a metric to monitor and evaluate their success [41, 42].

Reach refers to the number of people who have been exposed to a particular message or campaign. Understanding reach is important because it allows organisations to assess the extent of their impact and determine how successful they have been in reaching their target audience. By analysing reach, environmental communication campaigns can better understand the level of engagement and awareness generated by their efforts.

Monitoring reach can be done through various methods, such as tracking website traffic, social media engagement, and media coverage. These quantitative data points provide valuable insights into the effectiveness of the campaign and help organisations gauge the level of interest and participation from the audience.

Moreover, reach can also be used to evaluate the overall impact of the campaign on behaviour change and attitude shift. By monitoring reach, organisations can assess whether their message has been effectively communicated and whether it has influenced individuals to take action or make changes in their daily lives.

For example, if a campaign aimed at reducing plastic usage has a high reach but low engagement, it may indicate that the message is not resonating with the audience or that the call to action is not clear enough. By analysing reach in conjunction with other metrics, organisations can gain a comprehensive understanding of the campaign's effectiveness and make informed decisions on how to improve future efforts.

8.3.7 *Impact on Environmental Outcomes*

Monitoring and evaluating environmental communication campaigns based on their impact on environmental outcomes can provide valuable insights into whether the desired behaviour change is occurring as a result of the campaign [43–45]. For example, if a campaign is aimed at reducing plastic waste, the metric of monitoring the amount of plastic waste in landfills before and after the campaign can provide concrete evidence of the campaign's effectiveness. If there is a significant decrease in the amount of plastic waste after the campaign, it can be inferred that the campaign has been successful in influencing behaviour change.

Furthermore, using environmental outcomes as a metric for evaluating environmental communication campaigns can help in assessing the long-term effectiveness of these campaigns. By tracking environmental indicators such as air and water quality, biodiversity, and carbon emissions over time, it is possible to determine whether the positive changes resulting from the campaigns are sustainable and lasting. This can help in refining future campaigns and strategies to ensure continued progress in environmental protection.

Additionally, monitoring and evaluating environmental communication campaigns based on their impact on environmental outcomes can also help in holding organisations and governments accountable for their environmental responsibilities. By linking campaign goals to specific environmental indicators, it becomes easier to determine whether the promises made by these entities are being fulfilled. This can help in building trust with the public and ensuring transparency and accountability in environmental stewardship.

8.4 Conclusion

Monitoring and evaluating the impact of environmental communication strategies is crucial for ensuring the effectiveness of conservation efforts. By tracking and assessing the outcomes of communication initiatives, it is possible to make informed decisions on how to best reach the target audience and inspire action. The theoretical metrics outlined in this chapter provide a framework for assessing the impact of communication efforts and make informed decisions to improve future campaigns. These theoretical metrics include the information deficit model, the participatory communication model, stage of change model, social marketing model, diffusion of innovations theory, health belief model, agenda-setting theory, and framing theory. Additionally, the conceptual metrics discussed in this chapter offer a perspective on the impact of environmental communication efforts. By considering factors such as awareness, reach, engagement, behaviour change attitude change, and policy influence, environmental communication organisations can gain a more holistic understanding of how our messages are resonating with our audience and driving positive change. By utilising both theoretical and conceptual metrics, environmental communication campaigns can ensure that their communication efforts are making a difference in protecting our planet for future generations.

Recap
Environmental communication programmes are essential tools in the fight against climate change and environmental degradation. These initiatives aim to raise awareness about various environmental issues, from pollution to biodiversity loss, while encouraging individuals to adopt more sustainable behaviours. However, merely launching these campaigns is insufficient. It is equally crucial to monitor and evaluate their effectiveness to understand their impact and make necessary adjustments for better results. This can be done using different theoretical and conceptual metrics. Through continual improvement, such programmes can effectively contribute to a more sustainable future.

Questions
- What is your understanding of the concept of monitoring and evaluation of the impact of environmental communication strategies?
- What is the importance of monitoring and evaluation of environmental communication programmes?
- What are some of the theoretical metrics for monitoring and evaluating environmental communication programmes?
- What are some conceptual metrics that can be used to monitor and evaluate environmental communication programmes?

References

1. Arts EJ. EIA follow-up: on the role of ex post evaluation in environmental impact assessment. Groningen: Geo Press; 1998.
2. Azcárate J, Balfors B, Bring A, Destouni G. Strategic environmental assessment and monitoring: Arctic key gaps and bridging pathways. Environ Res Lett. 2013;8(4):044033. https://doi.org/10.1088/1748-9326/8/4/044033.
3. Carton L, Ache P. Citizen-sensor-networks to confront government decision-makers: two lessons from the Netherlands. J Environ Manag. 2017;196:234–51.
4. Cherp A, Partidário MR, Arts J. From formulation to implementation: strengthening SEA through follow-up. In: Handbook of strategic environmental assessment. Routledge; 2012. p. 515–34.
5. Sadler B, Dalal-Clayton B. Strategic environmental assessment: a sourcebook and reference guide to international experience. Routledge; 2012.
6. EC. Directive 2001/42/EC of the European Parliament and of the Council of 27 June 2001 on the assessment of certain plans and programmes on the environment. OJEC; 2001.

7. čavoški A. REFIT of the SEA Directive. Environ Law Rev. 2020;22(2):108–18. https://ec.europa.eu/environment/eia/sea-refit.html
8. ECGF. Compilation of CJEU case law on member states' obligation to remedy failures to carry out environmental assessments. European Commission: DG ENV – Environmental Compliance and Governance Forum; 2021. https://circabc.europa.eu/ui/group/cafdbfbb-a3b9-42d8-b3c9-05e8f2c6a6fe/library/cd912549-8639-4591-aaf5-f5f250a465f3/details
9. EPA. Review of effectiveness of SEA in Ireland. Ireland: Environmental Protection Agency; 2012. https://www.epa.ie/publications/monitoring--assessment/assessment/SEA-EFFECTIVENESS-REVIEW-MAIN-REPORT-2012.pdf.
10. EPA. Second review of strategic environmental assessment effectiveness in Ireland. Ireland: Environmental Protection Agency; 2018. https://www.epa.ie/publications/monitoring--assessment/assessment/second-review-of-sea-effectiveness-in-ireland.php.
11. EPA. Tiering of environmental assessment—the influence of SEA on project level EIA. Ireland: Environmental Protection Agency; 2021. https://www.epa.ie/publications/research/epa-research-2030-reports/Research_Report_391.pdf
12. EU environmental assessment of projects and plans and programmes—rulings of the Court of Justice of the European Union. Luxembourg: Publications Office of the European Union; 2020. https://ec.europa.eu/environment/eia/pdf/EIA_rulings_web.pdf.
13. Fischer TB, González A, editors. Handbook on strategic environmental assessment. Cheltenham (UK): Edward Elgar Publishing; 2021.
14. Gachechiladze M, Noble BF, Bitter BW. Following-up in strategic environmental assessment: a case study of 20-year forest management planning in Saskatchewan, Canada. Impact Assess Proj Apprais. 2009;27(1):45–56. https://doi.org/10.3152/146155109X430362.
15. Gachechiladze-Bozhesku M, Fischer TB. Benefits of and barriers to SEA follow-up—theory and practice. EIA Rev. 2012;34:22–30. https://doi.org/10.1016/j.eiar.2011.11.006.
16. González A, Bullock C, Gaughran A, Watkin-Bourne K. Towards a better understanding of SEA effectiveness in Ireland. IAPA. 2019;37(3–4):233–46. https://doi.org/10.1080/14615517.2019.1580475.
17. González Del Campo A, Gazzola P. Untapping the potential of technological advancements in Strategic Environmental Assessment. J Environ Plan Manag. 2020;63(4):585–603.
18. González A, Therivel R. Raising the game in environmental assessment: insights from tiering practice. EIA Rev. 2022;92:106695. https://doi.org/10.1016/j.eiar.2021.106695.
19. Gutierrez M, Bekessy SA, Gordon A. Biodiversity and ecosystem services in strategic environmental assessment: an evaluation of six Australian cases. EIA Rev. 2021;87:106552. https://doi.org/10.1016/j.eiar.2021.106552.
20. Hadi SP, Purnaweni H, Prabawani B. The power of Strategic Environmental Assessment (SEA): a case studies of North Kendeng Mountain Area, Central Java, Indonesia. In: E3S web of conferences, vol. 125. EDP Sciences; 2019. p. 02014.
21. Hanusch M, Glasson J. Much ado about SEA/SA monitoring: the performance of English Regional Spatial Strategies, and some German comparisons. EIA Rev. 2008;28(8):601–17. https://doi.org/10.1016/j.eiar.2007.12.001.

22. IAIA. Strategic environmental assessment: performance criteria. Fargo (USA): International Association for Impact Assessment; 2020.
23. Jiricka-Pürrer A, Wanner A, Hainz-Renetzeder C. Who cares? Don't underestimate the values of SEA monitoring! EIA Rev. 2021;90:106610.
24. Lamorgese L, Geneletti D, Partidario MR. Reviewing strategic environmental assessment practice in the oil and gas sector. J Environ Assess Pol Manag. 2015;17(02):1550017. https://doi.org/10.1142/S1464333215500179.
25. Lundberg K, Balfors B, Folkeson L, Nilsson M. SEA monitoring in Swedish regional transport infrastructure plans—improvement opportunities identified in practical experience. EIA Rev. 2010;30(6):400–6. https://doi.org/10.1016/j.eiar.2009.12.002.
26. Mascarenhas A, Ramos TB, Nunes L. Developing an integrated approach for the strategic monitoring of regional spatial plans. LUP. 2012;29(3):641–51. https://doi.org/10.1016/j.landusepol.2011.10.006.
27. Morrison-Saunders A, Arts J. Assessing impact: handbook of EIA and SEA follow-up. London: Routledge; 2004.
28. Nilsson M, Wiklund H, Finnveden G, et al. Analytical framework and tool kit for SEA follow-up. EIA Rev. 2009;29(3):186–99.
29. Partidário MR, Arts J. Exploring the concept of strategic environmental assessment follow-up. IAPA. 2005;23(3):246–57. https://doi.org/10.3152/147154605781765481.
30. Partidario MR, Fischer TB. Follow-up in current SEA understanding. In: Assessing impact. Routledge; 2012. p. 224–47.
31. Persson Å, Nilsson M. Towards a framework for SEA follow-up: theoretical issues and lessons from policy evaluation. J Environ Assess Pol Manag. 2007;9(04):473–96. https://doi.org/10.1142/S1464333207002901.
32. Polido A, João E, Ramos TB. Strategic Environmental Assessment practices in European small islands: insights from Azores and Orkney islands. EIA Rev. 2016;57:18–30. https://doi.org/10.1016/j.eiar.2015.11.003.
33. Sadler B. On evaluating the success of EIA and SEA. In: Assessing impact. Routledge; 2012. p. 248–85.
34. Sadler B, Dusik J, Fischer T, Partidario M, Verheem R, Aschemann R. Handbook of strategic environmental assessment. London: Earthscan; 2012.
35. Söderman T, Kallio T. Strategic environmental assessment in Finland: an evaluation of the SEA Act application. J Environ Assess Pol Manag. 2009;11(01):1–28. https://doi.org/10.1142/S1464333209003269.
36. Stepenuck KF, Green LT. Individual-and community-level impacts of volunteer environmental monitoring: a synthesis of peer-reviewed literature. Ecol Soc. 2015;20(3) https://doi.org/10.5751/ES-07329-200319.
37. Therivel R, González A. Introducing SEA effectiveness. IAPA. 2019;37(3–4):181–7. https://doi.org/10.1080/14615517.2019.1601432.
38. Therivel R, González A. "Ripe for decision": Tiering in environmental assessment. EIA Rev. 2021;87:106520. https://doi.org/10.1016/j.eiar.2020.106520.
39. Wallgren O, Nilsson M, Jonsson DK, Wiklund H. Confronting SEA with real planning: the case of follow-up in regional plans and programmes in Sweden. J Environ Assess Pol Manag. 2011;13(02):229–50.
40. Hermans T. Translation and history: a textbook. Routledge; 2022.

41. Hicks D, Bord A. Learning about global issues: why most educators only make things worse. Environ Ed Res. 2001;7(4):413–25. https://doi.org/10.1080/13504620120081287.
42. Jung CD, Dos Santos LM. Incorporating environmentally-responsive EFL pedagogy in English-as-a-foreign language classrooms: paving the way for global impact. In: IOP conference series: earth and environmental science, vol. 987. IOP Publishing; 2022. p. 012017. https://doi.org/10.1088/1755-1315/987/1/012017.
43. Kärnä J, Hansen E, Juslin H. Social responsibility in environmental marketing planning. Eur J Mark. 2003;37(5/6):848–71. https://doi.org/10.1108/03090560310465170.
44. Kassas M. Environmental education: biodiversity. Environment. 2002;22:345–51.
45. Katz-Kimchi M, Goodwin B. Organizing and integrating knowledge about environmental communication. Environ Commun. 2015;9(3):367–9. https://doi.org/10.1080/17524032.2015.1042985.

CHAPTER 9

Conclusions, Policy, Recommendations, and Future Direction to Overcome Environmental Communication Challenges

Introduction
Africa is experiencing rapid change, and environmental concerns are becoming more prevalent and significant than before. However, one of the biggest challenges in addressing these issues is the lack of effective communication between policymakers, scientists, and the public. Implementing policies and recommendations that support increased public engagement and communication is essential to effectively address these environmental communication challenges. This chapter focuses on policies and recommendations that can be implemented in order to enhance public engagement and communication regarding environmental concerns. The first part of the chapter discusses policies, followed by recommendations, and ends with a discussion on future directions of sustainable environmental communication.

9.1 Conclusions

This section synthesises the key points discussed throughout the text, inspiring further research and action, and driving positive change. By reflecting on the lessons learned and outlining future directions, the conclusions reinforce the importance of the topic and highlight the potential implications for policymaking, advocacy, and community engagement in environmental communication.

9.1.1 Concept of Environmental Communication

Effective environmental communication is crucial for raising awareness and inspiring people to take proactive measures to protect our planet [1]. By

© The Author(s), under exclusive license to Springer Nature Switzerland AG 2025
G. Mutinta, *An African Perspective on Environmental Communication*,
https://doi.org/10.1007/978-3-032-05052-6_9

engaging in the process of identifying, understanding the historical background, and setting clear objectives for environmental communication, environmental communication organisations may effectively engage with their target audiences to promote positive change. Technological improvements and variations in societal perspectives have exerted significant effect on the growth of environmental communication. These elements have had a profound influence on the way individuals approach environmental messaging. Environmental communication organisations may improve their relationship with the target audience and encourage meaningful action by applying effective communication concepts, such as transparency, storytelling, and captivating visuals. Environmental communication aims to empower individuals, communities, and society to take responsibility for their impact on the environment and work towards sustainable solutions. By employing efficient communication, environmental communication programmes can initiate significant changes in all areas and promote a greater environmental consciousness.

9.1.2 *Environmental Challenges*

Africa is facing several environmental challenges that require urgent attention and collaborative efforts to protect the well-being of its ecosystems and residents [2]. The issues of deforestation, land degradation, urbanisation, water shortages, and other factors are posing a threat not just to the biodiversity and ecosystems of the region but also to the livelihoods and health of millions of individuals. Collaboration among governments, non-governmental organisations, corporations, and people is crucial in addressing these challenges and implementing sustainable solutions. Examples of interventions that can help mitigate the effects of environmental degradation in Africa include conservation programmes, afforestation projects, sustainable land management practices, and effective water resource management strategies. Partnership between environmental communication organisations and stakeholders is vital to protect the environment, preserve biodiversity, guarantee the accessibility of water resources, and improve the general quality of life for citizens. Emphasising environmental communication is essential for enhancing the environmental condition and the welfare of its inhabitants, yielding advantages for both current and future generations.

9.1.3 *The Role of Traditional and Digital Media in Environmental Communication*

The media has a crucial role in environmental communication since it greatly contributes to raising awareness and supporting positive changes [3]. Traditional media channels should be employed in environmental communication since they have functioned as a medium for spreading news and information regarding environmental issues, successfully reaching a wide range of people and encouraging discussions. Due to the ongoing progress of technology, digital media has become increasingly important as a vital tool for environmental communication. It facilitates the dissemination of engaging and immersive content to a global audience. Media outlets, including films, social media campaigns, and Internet platforms, have proven to be effective in communicating environmental issues and influencing public opinion, leading to increased motivation for action. Environmental organisations and activists can effectively communicate the seriousness of environmental issues and inspire others to protect our planet by utilising traditional and digital media channels. Both traditional and digital media exert a substantial influence on altering public attitudes and actions with regards to environmental concerns. Hence, it is imperative for media professionals, environmental activists, and the general public to work together in order to use the power of media in advocating for advantageous changes that will contribute to a sustainable future.

9.1.4 *Traditional African Communication Tools*

The findings show the undeniable potential of traditional African communication methodologies in the field of environmental communication. Harnessing the power of oral tradition and folklore can significantly enhance community awareness and promote the adoption of sustainable practices. By utilising these established traditions, health communication can foster a deep connection between persons and the environment, ultimately leading to a harmonious and tranquil relationship with the natural world. Environmental communication organisations can successfully convey vital information about the importance of conservation and the urgent need to protect the environment by using storytelling and sharing folklore. These traditional methods of communication have a unique ability to engage and inspire listeners in a way that is both meaningful and impactful. In light of the pressing environmental challenges that Africa is presently confronting, it is imperative to employ traditional African communication tactics to bolster and advocate for sustainability and conservation projects. Environmental communication organisations can play a role in creating a sustainable future for everyone by adopting and supporting these important cultural communication practices.

9.1.5 *Challenges and Opportunities for Environmental Communication*

The challenges facing environmental communication in Africa are significant, but the opportunities for progress and success are equally strong [3–5]. Effective cooperation between governments, non-governmental organisations, and individuals is essential in addressing issues such as restricted information availability, language barriers, and opposing interests. By directing resources towards education and training, encouraging open conversations and collaboration, and utilising innovative technologies and platforms, environmental communication can successfully tackle these difficulties and provide the foundation for a more sustainable future on the continent. Africa possesses significant potential for environmental communication, since it can draw lessons from past mistakes and readily embrace cutting-edge solutions. Environmental communication organisations can effectively engage a wider audience and inspire action on a larger scale by highlighting the interconnectedness of environmental concerns with social, economic, and political factors. Environmental communication organisations can empower communities to take responsibility for their natural resources, advocate for policy reform, and foster a culture of environmental stewardship by employing successful strategies and building strong partnerships. Taking advantage of the opportunity and employing effective communication are essential for promoting positive environmental change in Africa.

9.1.6 The Role of Stakeholders in Environmental Communication

Stakeholders have a crucial role in environmental communication [6–9]. Their active engagement and dedication can greatly impact the efficacy of communication efforts aimed at addressing environmental issues. By recognising the importance of stakeholders and actively engaging them in the communication process, organisations and individuals can develop stronger ties, improve transparency, and ultimately achieve more meaningful results in their environmental communication initiatives. Stakeholders consist of a wide array of persons and groups, such as community members, government officials, and non-profit organisations. Each group brings forward their own perspectives, specialised expertise, and individual preferences, underscoring the need of considering all stakeholders when developing communication strategies. Environmental communicators can enhance the pertinence, impact, and efficacy of their communications by actively involving diverse communities. By recognising the importance of stakeholders and actively including them in the communication process, businesses and people may work towards a more sustainable and environmentally conscious future. Environmental communication organisations should prioritise stakeholder engagement in their efforts to tackle environmental issues. Collaboration and cooperation are important for these organisations to exert a significant influence and manage conflicts of interests among stakeholders. This requires transparency, effective communication, and a commitment to finding solutions that benefit the environment and all stakeholders. By working together collaboratively and considering the long-term impacts of their decisions, stakeholders can overcome conflicts of interests and work towards a more sustainable future.

9.1.7 The Role of Environmental Communication in Education

Environmental communication is unquestionably crucial in education as it significantly enhances awareness of environmental issues and promotes the concept of sustainability [10]. Highlighting the importance of environmental protection and conservation enhances the probability of persons participating in efforts to preserve the environment. Environmental organisations can engage the public in discussions on pressing environmental matters through the implementation of communication strategies, including campaigns, educational programmes, and lobbying activities. This can lead to increased awareness and also inspire individuals to make educated decisions that support environmental well-being. Furthermore, integrating environmental communication into academic courses can help foster a deep understanding of the interconnectedness between human activities and the environment. By incorporating sustainability concepts into their education, learners can be motivated to become advocates for the environment and catalysts for positive change. Environmental communication in education can be a powerful tool that can inspire collective action in order to create a more sustainable and resilient future for the planet. Emphasising environmental communication in schools is essential for cultivating a cohort of environmentally conscious individuals who can actively participate in the creation of a more sustainable and healthier planet.

9.2 The Importance of Policy in Environmental Communication

Environmental communication is crucial to advancing sustainable practices and raising public awareness of environmental issues [11–13]. In Africa, where the repercussions of environmental deterioration are becoming increasingly apparent, effective communication policies are even more crucial. Adequate policies for environmental communication are essential to ensure that messages are effectively communicated and that environmental problems are addressed.

One of the reasons that good environmental communication policies are essential is the complexity and diversity of environmental concerns that exist throughout the African continent. These issues, which range from pollution and water shortages to deforestation and desertification, call for understandable and all-encompassing communication policies. Effective policies ensure that messages are tailored to different audiences while taking language, cultural, and socioeconomic differences into consideration.

Effective environmental communication policies can also foster connections among stakeholders that are reliable and trustworthy. In Africa, where there is often a high degree of suspicion towards institutions of authority and organisations, transparent and inclusive communication techniques can help to foster cooperation and collaboration. Through meaningful interaction with communities, governments, and civil society groups, environmental communication techniques have the potential to promote collective action and a shared understanding of environmental challenges.

Moreover, resource generation and promotion for environmental initiatives can be achieved through efficient environmental communication policies. In Africa, where funding for environmental projects is usually scarce, effective communication policies can help to draw in support from organisations, the general public, and funders. By highlighting the benefits of environmental preservation for public health, the economy, and society as a whole, policies in environmental communication can bolster the case for sustainable development.

In addition, the implementation of efficient environmental communication policies is necessary to hold governments and corporations accountable for their environmental impacts. Many African countries have seen widespread environmental deterioration as a result of opaque decision-making processes and a dearth of enforcement of environmental regulations. By establishing explicit communication guidelines and monitoring and assessment processes, policies in environmental communication can help ensure that stakeholders are held accountable for their actions and that progress towards environmental goals is monitored.

9.2.1 *Tailoring Messaging to Local Realities*

With its diverse population of languages, traditions, and religious beliefs, Africa is a cosmopolitan continent [14]. Communication that is tailored to the local environment is more reliable and effective. When environmental messages are communicated using relatable language and symbols, they are more likely to be understood and adopted by the intended audience. Through the process of bridging the information gap between scientific and local sources, better collaboration and communication can be fostered for environmental protection in Africa.

Respecting and upholding traditional knowledge and practices is made simpler when environmental communication considers local circumstances. Many African communities have strong ties to the environment and can provide valuable advice on living sustainably. Acknowledging and incorporating diverse points of view into communications can enhance inclusivity and efficiency in environmental communication. Tailoring messaging to local realities can also give local populations a sense of pride and ownership, motivating them to participate in conservation efforts.

Messages can be tailored to specific environmental challenges and concerns in different African locales by taking into account local settings. Community-specific environmental issues include conservation of animals, deforestation, and water scarcity. By adjusting communications to various regional issues, environmental communication can efficiently organise resources and support for certain solutions. This tactic can raise the possibility of a successful change by ensuring that attention is focused on the most pressing issues in Africa.

When environmental messages are customised for local conditions, they acquire trust and confidence. Communications that align with the values and practical experiences of a community are more likely to be accepted by that group. Communicators who show that they are aware of local contexts and challenges can build stronger links and more trusted relationships with communities. This creates opportunities for important coalitions and cooperation, both of which are essential for long-term environmental impact in Africa.

9.2.2 *Using the Traditional Channels of Communication*

In Africa, there is a huge digital divide, with many people living in remote areas without access to Internet or telephones [15]. This suggests that traditional communication channels like radio, television, print media, oral tradition, folklore, and community meetings are occasionally the only ones accessible for certain demographic groups, such as the elderly and those living in remote regions. By using these channels, environmental messages might potentially reach a wider audience, ensuring that everyone is included in the fight against environmental degradation and climate change.

Traditional communication channels are considered reliable sources of information in African communities since they have been there for a long time. People look on their local radio stations and leaders in the community to provide them with timely and reliable information. By utilising these channels, environmental organisations can improve their standing and credibility in the community, which can raise the possibility that people will hear their messages and take action.

Traditional communication techniques, which are tailored to the particular needs and preferences of the community, are deeply rooted in African culture. For example, radio plays and storytelling are prevalent forms of communication in many African countries. By tailoring their communications to the cultural values and beliefs of the community, environmental groups can utilise these platforms to increase the impact and engagement of their messages.

Conventional means of communication can be used to reinforce and magnify environmental signals. Environmental organisations can ensure that their ideas are understood by using a multi-channel approach that blends traditional media with online channels. For instance, a radio programme on the importance of sustainable farming methods can be enhanced by social media updates that highlight the achievements of nearby farmers. Utilising several channels increases the likelihood that the message will be heard, understood, and acted upon.

Collaborating with grassroots campaigns, civic leaders, and neighbourhood groups to advance environmental message and initiatives is one of the good strategies for environmental communication. By including significant stakeholders in communication campaigns, environmental communication can boost trust and credibility in African local communities.

9.2.3 *Emphasising the Benefits of Sustainability*

Stressing the benefits of sustainability in African environmental communication can help to raise awareness of the need to protect the natural resources that are essential to the continent. Africa is home to some of the world's most biodiverse ecosystems, which offer vital services including rich soil, clean air, and clean water. By highlighting the benefits of sustainability, environmental communication can help protect these resources for future generations.

Underscoring the benefits of sustainability in environmental communication might help address some of the major environmental issues that Africa is now experiencing. A number of environmental problems that Africa is dealing with all of which threaten the livelihoods of millions of people. To mitigate the consequences of environmental issues and build a more sustainable future for Africa, environmental communication should promote sustainable activities like reforestation, soil conservation, and renewable energy.

In African environmental communication, highlighting the benefits of sustainability can help reduce poverty and promote economic progress [16]. Organic farming, eco-tourism, and renewable energy are examples of sustainable businesses that improve local communities by creating jobs and income. These kinds of businesses can help lift people out of poverty and improve their quality of life. By highlighting the financial benefits of sustainability, environmental communication can influence businesses, individuals, and organisations to adopt more environmentally friendly activities.

Highlighting the benefits of sustainability in environmental communication in Africa can also help build resilience in the face of climate change and other environmental concerns. Supporting sustainable practices like resource conservation, ecological restoration, and disaster preparedness can help communities become more resilient to the consequences of natural catastrophes and other challenges. Sustainability is not just protecting the environment but also ensuring human well-being and long-term existence in a changing world.

9.2.4 Promoting Two-Way Communication

It is possible to have a more thorough understanding of environmental concerns through reciprocal communication [17]. Engaging in dialogue with a diverse array of stakeholders, including companies, non-governmental organisations, government agencies, and local communities, can enhance comprehension of Africa's environmental issues. Consequently, this can enable the construction of more robust and effective environmental solutions.

Two-way communication encourages empowerment and inclusivity. Ensuring inclusivity of all viewpoints in environmental decision-making can give marginalised communities a platform to voice concerns and recommendations. This can ensure that their needs are satisfied and provide them with the self-assurance to play a more active role in shaping their own environments and futures.

Besides, two-way communication can improve accountability and transparency. By having open and honest conversations with interested people, governments and organisations can build credibility. Consequently, this can ensure that environmental regulations are implemented equitably and efficiently and contribute to holding decision-makers accountable for their decisions.

Two-way communication facilitates responsive and adaptable environmental management. In an African environment that is changing rapidly, it is imperative that environmental laws and practices be adaptable enough to take into account new opportunities and issues. By maintaining constant contact with interested parties, policymakers can stay informed about emerging issues and adjust their approaches accordingly.

9.2.5 *Utilising Digital Technologies*

Digital technologies offer a cost-effective and efficient way to disseminate environmental information. Given how common cellphones and Internet access are throughout Africa, digital platforms provide a quick and effective means of reaching a large audience [18]. Environmental communication organisations should use social media, websites, and mobile applications to communicate with a wide range of stakeholders and raise awareness of pressing environmental concerns.

Digital technologies enable data collection and real-time monitoring, both of which are necessary for effective environmental management. Using satellite imagery, drones, and sensors, environmental scientists can investigate changes in ecosystems, track wildlife populations, and assess how human activity affects the environment. These technological advancements can provide informative data that can impact African decision-making and conservation programmes.

Besides, digital technologies facilitate collaboration and knowledge-sharing among stakeholders in the environmental activities. Virtual spaces and online platforms can allow people to communicate, exchange ideas, and collaborate on environmental preservation efforts. This Internet connectivity in Africa can foster a sense of community and enable individuals to collaborate for environmental sustainability.

In addition, digital technology gives people the ability to take charge and transform their communities. Ordinary people can use online campaigns, petitions, and crowdfunding platforms to mobilise support for environmental issues and contribute to conservation initiatives. Using modern tools, people can raise awareness, advocate for positive improvements to Africa's environmental predicament, and magnify their voices.

9.2.6 *Building Capacity and Skills*

Effective environmental communication is essential to promoting understanding and increasing awareness of environmental issues. Africa is a continent abundant in natural resources and wildlife, but it also faces several environmental challenges. By giving individuals and organisations the resources they need to effectively communicate these issues, environmental communication can increase public awareness and encourage support for environmental conservation efforts.

Increasing environmental communication skills can help with stakeholder connections and collaboration. To tackle complex environmental concerns, cooperation between several organisations—including politicians, scientists, and local populations—is essential. Effective communication techniques can encourage the collaboration and dialogue required to provide long-lasting environmental solutions by bridging the gap between these groups.

Acquiring proficiency in environmental communication can aid local communities' efforts to advance environmental conservation and sustainable development. In many African countries, the poor are disproportionately affected by environmental degradation and climate change. By providing these communities with the ability to voice their concerns and share their knowledge, environmental communication can help them speak up more and gain greater clout in influencing environmental practices and policy.

Building environmental communication skills can help Africa achieve sustainable development and economic growth. The increasing urbanisation and industrialisation of the continent has made effective communication about green technologies and sustainable practices increasingly crucial. Providing individuals with the knowledge and communication abilities they need to discuss these issues can help to attract investment, encourage innovation, and create new opportunities for green businesses and employment.

9.2.7 Promoting Storytelling and Visuals

Stories and visuals have the power to emotionally engage and connect people [19]. When information is presented in the form of an interesting story or visually striking picture, it just cannot captivate people in the same way as raw numbers and figures. This emotional connection can make people more understanding and informed about environmental issues, which can motivate them to take action and improve their own lives.

Compelling stories and visual aids can make complex scientific concepts easier to grasp for a wider audience. Environmental issues can be extremely complicated and difficult for the average individual to understand. Telling stories and using visual aids can make these issues more relatable and easier for people to understand in their daily lives. As a result, people could be in a better position to promote sustainable practices in their communities and make wise decisions.

Furthermore, stories and visuals can highlight the ways in which environmental issues are linked to social, economic, and cultural factors. Raising awareness of the potential wide-ranging repercussions of environmental degradation that can extend beyond the physical environment is essential, particularly in Africa where a large number of people rely on natural resources for their livelihoods. By emphasising the human stories behind these difficulties, environmental communication can encourage collaboration across sectors and foster a sense of shared responsibility in order to properly solve these concerns.

Using story and imagery in environmental communication can help to highlight the viewpoints of poor populations which are disproportionately affected by environmental deterioration. By giving these groups, a platform to express their stories and experiences, people can draw attention to the injustices and inequities that lie at the root of many environmental issues. This can lead to a rise in public awareness, activism, and support for policies that prioritise environmental justice and sustainable development.

9.2.8 Promoting Collaboration and Networking

Fostering collaboration and networking enable the sharing of resources and expertise. By working together, individuals and organisations in Africa can pool their resources and knowledge to develop creative solutions to environmental issues. As a result, solutions to issues like deforestation, water scarcity, and animal protection can become more realistic and effective. Through collaboration, African countries can leverage each other's expertise and combine their resources to tackle environmental issues more successfully.

Collaboration and networking can help champions of the environment get their voice heard. By working together, individuals and organisations can multiply their messages and reach a wider audience. This can help raise public awareness of important environmental issues and boost support for conservation efforts. By banding together and networking, environmental activists can become more well-known and influential, which can make it easier to promote sustainable practices and change laws.

In Africa, collaboration and networking can foster a sense of unity and mutual support among environmental activists. By working together to accomplish a similar goal, individuals and organisations can forge strong relationships and partnerships that transcend boundaries and cultural barriers. As a result of this, environmentalists could feel more unified and prepared to work together to achieve a shared objective: a sustainable future. By encouraging networking and collaboration, African countries can work together to address environmental concerns and build a more resilient and sustainable continent.

Collaboration and networking can help bridge the gap between scientific research and policymaking. Together, scientists, policymakers, and advocacy groups can collaborate on research projects and share knowledge. This can help to inform evidence-based policymaking and ensure that decisions are founded on scientific research and best practices. By promoting networking and collaboration, African countries can enhance their capacity to make evidence-based choices and ensure that environmental policies are founded on the most recent data.

9.2.9 Monitoring and Evaluating Impacts

It is possible to monitor and evaluate environmental communication strategies in order to determine their effectiveness. People can monitor important metrics like reach, engagement, and behaviour change to find out what is working and what needs to be improved. As a result of data-driven approach, people can be able to make well-informed decisions on the best way to allocate resources and improve message for maximum impact.

Monitoring and assessing the impact of environmental communication can help keep parties accountable. Particularly in Africa, where environmental challenges are often intertwined with social, economic, and political concerns, it is imperative to ensure that communication activities are not just reaching the intended audience but also yielding observable outcomes. When tracking the development of environmental communication programmes and document successes—or failures—funders, policymakers, and the general public can all see the advantages.

Impact monitoring and evaluation also can assist in learning from mistakes. In the constantly changing African continental of environmental communication, certain projects inevitably fall short of expectations. By doing in-depth evaluations, people can determine what went wrong, why it went wrong, and how to avoid making the same mistakes in the future. It would take a continuous upgrading effort to create a more reliable and efficient communication infrastructure throughout Africa.

Additionally, a responsible and open culture can be developed via impact monitoring and evaluation. In a field where environmental issues are often downplayed or ignored, it is imperative to emphasise the role that communication plays in encouraging positive change. By being open and honest about findings and including stakeholders in the evaluation process, it is possible to foster the type of collaboration and trust that are essential for long-term success.

9.3 Recommendations

Recommendations presented in this section provide valuable insights into how communication can be utilised to promote environmental conservation and sustainable development on the continent. These recommendations based on these findings are essential for guiding policymakers, environmental organisations, and other stakeholders in effectively communicating environmental issues to the public.

9.3.1 Government Agencies

Government agencies must communicate in an intelligible and clear manner when discussing environmental problems. Since many Africans might not be familiar with scientific concepts or technical jargon, it might help to guarantee that ideas are understood by a wide audience to communicate them in an acceptable way. By breaking down complex information into manageable portions, government agencies can empower people to make informed decisions regarding their environment.

Government agencies should employ a variety of communication channels to engage with different demographic groups. Even though African consumers are increasingly using digital platforms, traditional media like oral tradition, radio, and television still have a role to play. Social media, websites, and mobile applications are useful ways to reach younger generations and urban individuals who can be more likely to engage with environmental issues online. Governmental entities can use a variety of communication strategies to ensure that a broad audience hears their messages.

Governmental agencies should place a high premium on encouraging community participation and engagement in environmental communication programmes. Rather than merely spreading information, agencies should seek to create opportunities for dialogue and engagement with local populations. By including people in decision-making processes, agencies can foster trust, a sense of responsibility over environmental issues, and eventually more sustainable results. Promoting neighbourhood activities can lead to long-lasting improvements that benefit both people and the environment.

Encouraging and inspiring individuals to take up environmental issues should be the aim of government agencies. Even while it is important to emphasise how severe the issues confronting Africa are, a message of doom and gloom can often leave people feeling helpless or indifferent. Instead, agencies need to highlight success stories, show how both individual and collective efforts can yield positive results, and offer workable solutions that people can utilise on a daily basis. Governmental agencies that communicate environmental issues in a positive and effective way can encourage people to actively participate in environmental protection.

9.3.2 Policymakers

One of the most important aspects of environmental communication is educating the public about the importance of environmental sustainability and conservation. Policymakers in Africa should support targeted campaigns and initiatives that educate and engage the public on significant environmental issues including pollution, deforestation, and biodiversity loss. By increasing public awareness, policymakers can motivate people to take action to protect the environment by fostering a feeling of urgency and responsibility.

Effective environmental communication often requires collaboration and partnerships with local communities and non-governmental organisations. Policymakers in Africa have to work closely with non-governmental organisations and the local populace to develop and implement communication techniques that are culturally appropriate, responsive to local needs and values, and both. Through community and non-governmental groups, policymakers can ensure that their views are heard and understood by a larger audience during the communication process.

In the current digital era, social media and digital platforms are efficient ways to communicate with the public and share information. African officials ought to exploit these platforms to reach a larger audience with their environmental messaging. By creating engaging and informative content for social media and other digital platforms, policymakers can effectively mobilise support and raise awareness for environmental issues.

African policymakers must engage youth through education and awareness campaigns since they are the future generation of environmental stewards and leaders. By incorporating environmental education into the curriculum and organising campaigns and programmes with a youth emphasis, policymakers can empower young people to become environmental activists and change makers. By including children in environmental communication programmes, policymakers can inspire the next generation of environmental leaders and build a more sustainable future for Africa.

9.3.3 Media Organisations

One of the most effective ways for media organisations to raise public awareness of environmental issues is to incorporate environmental stories into their regular programming. Through news articles, documentaries, or talk shows, media outlets can educate their audience on the consequences of human activities on the environment and support local initiatives that are promoting sustainability. By including environmental issues into their regular programming, media organisations can foster a culture of environmental knowledge and action among their audience.

Media organisations can increase their reach and reputation by collaborating with African environmental experts and organisations. By partnering with non-governmental organisations, academic institutions, and thought leaders in the field, media outlets can get valuable insights and ideas that enhance their coverage of environmental issues. Through these collaborations, media outlets can promote environmental initiatives and campaigns while also involving their readers in thought-provoking conversations on sustainability and conservation.

Media outlets should focus on reporting on innovative solutions and heartwarming stories of environmental stewardship and resourcefulness, in addition to covering environmental concerns and threats. Media sources that highlight initiatives and success stories that are enhancing Africa can inspire their audience to take action and become involved in environmental conservation. People can be motivated to become change agents in their communities if the narrative around environmental problems shifts from one of helplessness to one of empowerment and solutions.

In the modern digital age, media organisations have a plethora of platforms at their disposal to engage with their audience on environmental issues. Through social media, podcasts, or interactive websites, media organisations can create interactive platforms that encourage conversations on environmental issues and provide viewers an opportunity to share their own opinions and experiences. By encouraging a dialogue with their audience, media organisations can build a community of environmentally conscious people who are committed to taking action to protect the environment.

Moreover, climate change, biodiversity loss, pollution, and human-induced factors are posing significant environmental challenges. Journalists play a crucial role in raising public awareness and informing policymakers about these issues. Therefore, there is a pressing need for training environmental journalists in Africa to ensure accurate, timely, and impactful reporting. This specialised field requires knowledge of scientific concepts, policy frameworks, and environmental issues. Training journalists can enable them to investigate and report on complex environmental stories accurately and effectively, raising public awareness, holding governments and corporations accountable, and prompting action to address environmental challenges. Environmental journalists can also promote sustainable development and protect natural resources, highlighting success stories and innovative solutions. Training can amplify these positive stories and create a ripple effect across Africa. Additionally, it can build a network of skilled journalists, promoting in-depth reporting, greater impact, and increased public engagement with environmental issues.

9.3.4 Civil Society Organisations

When talking about environmental issues, one must take into account the enormous diversity of African languages and cultures. Civil society organisations need to try to communicate in the languages that are spoken in the communities they are trying to engage with. By doing this, they can close the communication gap and ensure that messages are understood and resonate with the target audience.

Access to traditional media such as newspapers and television is limited in many African rural areas. Civil society organisations should employ a variety of communication channels, including social media, radio, and community events, to reach a wider audience. By using a range of channels, organisations can ensure that their messages are received by a diverse audience.

To effectively communicate about environmental concerns, civil society organisations must collaborate with local stakeholders, including elected officials, community leaders, and grassroots organisations. By working together, organisations can ensure that their messages are relevant, considerate of cultural variations, and reach the right audiences. Working with local stakeholders can help organisations gain the confidence and credibility of the community and improve support for their environmental programmes.

When communicating about environmental concerns, civil society organisations should focus on producing stories that resonate with audiences. By presenting stories of individuals or communities affected by environmental degradation, organisations can establish an emotional connection with their audience and encourage action. Stories that make environmental issues more relatable, pressing, and genuine have the power to motivate people to take action.

9.3.5 *Researchers*

In Africa, where environmental issues are becoming increasingly urgent, researchers have a unique opportunity to contribute valuable insights and solutions to some of the most pressing challenges facing the continent. Key research areas that environmental communication researchers in Africa should focus on include:

- **Indigenous perspectives and traditional knowledge:** Africa tribes are rich in knowledge and experience about the environment and sustainable living. Researchers should work to integrate and amplify these traditional viewpoints into the mainstream environmental discourse in order to ensure that a variety of voices and perspectives are heard and respected. By building a bridge between traditional knowledge and modern science, researchers can have the ability to improve mutual understanding and cooperation between diverse communities and promote more inclusive and effective environmental policies and practices.
- **Environmental justice and equity:** Among the marginalised and vulnerable groups that are disproportionately affected by many environmental issues in Africa are women, people of colour, and low-income communities. Researchers should investigate how communication processes shape perceptions of environmental benefits and risks, as well as how these perceptions affect the choices made and the outcomes of policies. By examining issues of power, privilege, and representation in environmental communication, researchers can help identify and address instances of environmental injustice while also working towards more equitable and sustainable solutions that benefit all members of society.
- **Public engagement and science communication:** Good communication has a big impact on how the public views and comprehends scientific research and environmental issues. Researchers operating in Africa should look into a range of communication strategies and platforms, such as social media, community radio, and traditional storytelling, in order to engage diverse audiences and foster discourse on environmental issues. By working with local communities and organisations, researchers can create persuasive communication campaigns and educational programmes that inspire people to take action and make responsible environmental decisions.
- **Climate change adaptation and resilience:** As the effects of climate change intensify, it is imperative that researchers examine how different African communities are managing the changes and getting ready for more climate-related challenges down the road. By examining communication practices and processes related to climate change adaptation, researchers can help identify and promote effective strategies for knowledge exchange, capacity-building, and community collaboration. By presenting best practices and exemplary case studies, academics can inspire

and inform others to take proactive steps and work together to build a more resilient and sustainable future for Africa.

9.3.6 *Educators*

One of the most effective ways to promote environmental communication is to incorporate environmental education into the curriculum. By educating learners about the importance of conservation, sustainable practices, and the effects of climate change, environmental communication can help them get a complete understanding of these issues and prepare them to become environmental advocates. To get learners thinking critically and involved in class debates, educators should create lesson plans that show real-world environmental concerns and their solutions.

Giving learners the tools they need to take action is another crucial aspect of environmental communication. Educators should support and encourage student-led initiatives such as tree planting drives, recycling efforts, and environmental awareness campaigns. By allowing learners to have a positive influence on their communities, environmental communication can help them develop a sense of agency and responsibility for environmental issues. Through these initiatives, learners can improve their communication abilities and engage with a wider audience on environmental concerns.

In the contemporary digital age, technology can be a useful tool for environmental communication. To engage learners in creative learning, educators might utilise social media, podcasts, and online forums to reach a wider audience. Educators can enhance their learners' environmental communication skills and accessibility by incorporating multimedia resources into their lessons. Utilising technology to facilitate cooperation, learners can also work together on environmental projects and share their thoughts with a global audience.

As educators, it is their responsibility to set a good example for their learners when it comes to environmental communication. Educators should set an example for others to follow by incorporating sustainable practices into their own lives and classrooms, such as reducing garbage, conserving energy, and promoting eco-friendly conduct. By leading by example, educators can inspire learners to adopt environmental stewardship in their local communities. Educators can also collaborate with other community members to strengthen their impact and create a more sustainable future for Africa, such as local non-governmental organisations and environmental groups.

9.3.7 *Community Leaders*

When talking about environmental problems in communication, it is crucial to include a range of community stakeholders. This includes local residents, the government, businesses, youth groups, and non-profit organisations. By including all relevant stakeholders in the conversation, community leaders can ensure that their messages are understood and lead to meaningful action.

Hearing the thoughts and perspectives of many groups can also stimulate original solutions to environmental issues.

To maximise impact and reach a wider audience, community leaders ought to employ a variety of communication channels. This might make use of both more traditional methods like radio broadcasts, meetings, and posters and more modern tools like social media, websites, and mobile applications. Using a range of media, community leaders can effectively connect with a variety of audiences and encourage more involvement in environmental projects.

Effective environmental communication in Africa requires teamwork. To maximise resources and expand their impact, community leaders should aggressively seek partnerships with other organisations, governmental agencies, and educational institutions. By working together to accomplish common goals, community leaders can strengthen their overall influence on environmental issues and build a strong network of support. Collaborating with external partners can enhance your decision-making process by contributing fresh perspectives and expertise.

Education is a powerful tool for promoting environmental-friendly behavioural modifications. Setting education and awareness-raising initiatives as a top priority can assist community leaders in providing residents with the knowledge and skills they need to protect the environment. Organising community events, training sessions, or seminars on topics like waste management, energy efficiency, and biodiversity preservation can fall under this category. By giving people in the community the information and resources they require, leaders can inspire people to take initiative and adopt the role of environmental stewards.

9.3.8 Sponsors of Environmental Communication

Funders ought to prioritise projects that include local communities and African grassroots organisations. Since these groups are often the most affected by environmental concerns, their unique views and skills can provide valuable insights for effective communication initiatives. By providing local voices and leaders more clout, funders can ensure that their efforts are relevant, culturally sensitive, and have a real impact on the ground.

Funders should allocate funds for innovative and multimodal communication strategies that engage audiences in novel and captivating ways. In the digital age, when attention spans are short and information overload is common, it is imperative that communication campaigns stand out and capture the interest of target audiences. Using a range of media, including movies, social media, infographics, and interactive tools, funders can effectively convey complex environmental concepts and encourage action.

Moreover, funders ought to prioritise the capacity-building and skill development of organisations and individuals engaged in environmental communication initiatives. Funders can help these initiatives become more effective and sustainable over the long run by providing resources, training, and mentorship.

This includes honing skills in storytelling, data analysis, stakeholder engagement, and communication strategy creation. By investing in capacity-building, funders can ensure that their efforts have a lasting impact and contribute to the strengthening of the African environmental champion's network.

Funders should place a high premium on cooperating and collaborating with other stakeholders, including governmental agencies, non-profit organisations, educational institutions, and the private sector. By pooling their resources, expertise, and networks to work towards a shared goal, funders can make environmental communication initiatives more effective. Collaborating can facilitate the dissemination of messages, encourage collective action towards sustainable solutions, and reach a wider audience. By fostering partnerships, funders can create a more cohesive and well-coordinated approach to addressing environmental concerns in Africa.

9.3.9 Curriculum Developers for Educational Institutions

Integrating environmental content into every topic should be a major priority for curriculum developers. This means adding relevant environmental material to science, geography, and social sciences courses, among other topics. Rather of being limited to one topic area, environmental education ought to be included throughout all of them. By doing this, learners can have a greater understanding of environmental issues and how they relate to other academic disciplines.

It is advised that those who develop curricula incorporate chances for experiential learning into their programmes. Since practical projects can help learners relate to environmental issues, experience truly is the greatest teacher. This can include organising field trips for learners to neighbouring parks or nature reserves, including them in outdoor education activities, or contributing to environmental conservation projects. By engaging in practical activities, learners will get a deeper understanding of the importance of talking about environmental concerns and a broader appreciation for the environment.

Curriculum development should place a strong emphasis on the development of communication skills. Effective communication is crucial to bringing about change and raising public awareness of environmental issues. Teaching learners how to write, speak in front of an audience, and use multimedia presentations to convey environmental concerns in an effective and elegant manner is crucial. If learners acquire these skills, they can be more equipped to advocate for positive change and debate environmental issues with their peers, teachers, and communities.

It is the duty of curriculum designers to provide opportunities for learners to engage in environmental advocacy and action. If learners are given a platform to address environmental issues, they have the potential to become powerful change agents in their communities. This can involve organising campaigns, participating in clean-up efforts, or collaborating on environmental projects with local organisations. By actively involving learners in environmental activism, curriculum designers can foster leadership and a sense of responsibility in the future generation of environmental communicators.

9.4 Future Direction

Effective communication on the environment is crucial in Africa to shape public opinion, legislative decisions, and ultimately the future of our planet. Africa's primary focus has to be on increasing public awareness and understanding of the serious issues including pollution, deforestation, climate change, and the loss of species.

It is imperative that indigenous knowledge and perspectives be given priority in environmental communication. Indigenous communities have deep environmental knowledge as well as perceptive knowledge that might direct sustainable practices and conservation. By giving their viewpoints more weight, this can encourage saner and workable solutions and increase understanding of environmental issues.

African countries must employ technology to enhance environmental communication throughout Africa. With the widespread use of mobile phones and Internet connectivity across the continent, people have an opportunity to reach a wider audience and start important dialogues about environmental conservation. Social media, digital platforms, and multimedia content should be employed to enable people to share compelling narratives that inspire others to take action and effect good change.

Encouraging collaboration and partnerships among all stakeholders engaged in environmental communication is crucial. Many different sources, such as local communities, media outlets, academic institutions, governments, and non-governmental organisations, have a significant impact on public discourse and policy choices. In the fight against environmental degradation, environmental communication can create a united front and increase our impact together.

Building capacity and providing training must come first in environmental communication. If journalists, communicators, and activists are equipped with the skills and expertise necessary to effectively and efficiently explain complex environmental problems, accurate information can be disseminated and understood by the general public. Investing in environmental communicators through training programmes, seminars, and tools can support each other in our efforts to raise awareness and spur action.

In environmental communication, it is important to adopt a solutions-focused approach. African countries must draw attention not only to the issues and threats facing the globe but also to the successes, innovations, and best practices that inspire and give us hope. By highlighting positive examples of sustainability and resilience, African countries can empower individuals and groups to effect autonomous change.

Prioritising inclusivity and cultural diversity is crucial in environmental communication. There are many different languages, cultures, and religious traditions throughout Africa, and each has its own unique perspective on the natural world. By recognising and celebrating diversity and incorporating cultural elements into communication strategies, environmental communication can create messages that resonate with individuals on a deeper, more intimate level.

African countries must foster a culture of environmental consciousness and social responsibility via communication initiatives. By bringing attention to the interconnectedness of all living things and common responsibility to protect the environment, African communities can foster a sense of togetherness and shared purpose among individuals and communities. This can help to have sustainable environmental communication to overcome the challenges presented by climate change and create a more resilient and sustainable future for everybody.

9.5 Conclusion

The importance of public engagement and communication regarding environmental concerns cannot be overstated. As the environment faces unprecedented challenges such as climate change and biodiversity loss, it is essential for policymakers, organisations, and individuals to work together to find solutions and take action. The policies discussed in this chapter provide a solid foundation for promoting public engagement and sustainable communication on environmental issues. By incorporating transparency, accountability, and inclusivity into decision-making processes, policymakers can build trust and support among the public. This, in turn, environmental communication initiative can lead to more effective environmental policies and actions. Furthermore, the recommendations outlined offer practical steps that stakeholders can take to improve sustainable communication and engagement efforts. From enhancing sustainable communication strategies to fostering partnerships with diverse audiences, there are numerous opportunities to enhance public understanding and support for sustainable environmental communication initiatives. Moving forward, it is imperative to prioritise sustainable environmental communication and engagement. By harnessing the power of sustainable environmental communication to educate, inform, persuade, and inspire action, it is possible to create a more sustainable future of the natural world.

Questions

- What conclusions about environmental communication stand out to you?
- What are some of the environmental communication policies that should be implemented?
- What are some of the environmental communication recommendations that should be implemented?
- What is your view of the future direction of environmental communication?

Recap
Africa is undergoing rapid transformations in its socioeconomic landscape, which have worsened the effects of environmental degradation and amplified the urgency of addressing environmental concerns. However, a major hurdle in tackling these issues lies in inadequate communication because there is a dearth of knowledge to inform environmental communication initiatives. To effectively confront the environmental challenges facing Africa, it is critical to establish a robust framework for public engagement and sustainable environmental communication using sound policies, recommendations, and future directions, as suggested in the chapter.

References

1. Abreu A, Pedrotti ML. Microplastics in the oceans: the solutions lie on land. Field actions science reports. J Field Actions. 2019;1(Special Issue 19):62–7.
2. Agustina A, Dewi TT, Soemantri NP, Qureshi NY, Moenanto G. Environmental communication in Indonesian television news coverage. Aspiration. 2020;1(1):56–72.
3. Suryawati GAI, Joni DAS. Strategi komunikasi Bye Bye Plastic Bags dalam Mengedukasi Masyarakat untuk Mengurangi Penggunaan Kantong Plastic [Bye Bye Plastic Bags communication strategy in educating the public to reduce the use of plastic bags]. Jurnal Ilmu Komunikasi [J Comm Sci]. 2021;1(2):1–6.
4. Andarsih YS, Mayangsari ID. Strategi Komunikasi Program Bandung Clean action dalam Mengkampanyekan Gerakan Pungut Sampah (GPS). eProceed Manag. 2017;4(2)
5. Anderson AA, Huntington HE. Social media, science, and attack discourse: how Twitter discussions of climate change use sarcasm and incivility. Sci Commun. 2017;39(5):598–620. https://doi.org/10.1177/1075547017735113.
6. Larasati A, Jamil A, Briandana R. Communication strategies in providing good government education through social media: a case study at the Parliament of the Republic of Indonesia. Psych Educ. 2021;58(1):722–34. https://doi.org/10.17762/pae.v58i1.822.
7. Ardian HY. Kajian teori komunikasi lingkungan dalam penelitian pengelolaan sumber daya alam. Jurnal Perspektif Komunikasi [J Commun Perspect]. 2018;2(1):1–20.
8. Aye L, Widjaya ER. Environmental and economic analyses of waste disposal options for traditional markets in Indonesia. WM. 2006;26(10):1180–91. https://doi.org/10.1016/j.wasman.2005.09.010.
9. Azmi NJ, Omar NA, Zaid SB, Wahab Z, Yusof A. Media portrayal on global climate change: an analysis of Malaysian mainstream newspapers. SMC. 2015;3(1):73–8. https://doi.org/10.11114/smc.v3i1.815.
10. Bentley AP, Petcovic HL, Cassidy DP. Development and validation of the anthropogenic climate change dissenter inventory. Environ Educ Res. 2019;25(6):867–82. https://doi.org/10.1080/13504622.2016.1250150.
11. Bolsen T, Shapiro MA. The US news media, polarization on climate change, and pathways to effective communication. Environ Commun. 2018;12(2):149–63. https://doi.org/10.1080/17524032.2017.1397039.

12. Briandana R, Doktoralina CM, Hassan SA, Hasan WN. Da'wah communication and social media: the interpretation of millennials in Southeast Asia. IJEBA. 2020;8(1):216–26. https://doi.org/10.35808/ijeba/543.
13. Carlson CJ, Albery GF, Merow C, et al. Climate change increases cross-species viral transmission risk. Nature. 2022;607(7919):555–62. https://doi.org/10.1038/s41586-022-04788-w.
14. Carter N, Childs M. Friends of the Earth as a policy entrepreneur: 'The Big Ask' campaign for a UK Climate Change Act. Environ Pollut. 2018;27(6):994–1013. https://doi.org/10.1080/09644016.2017.1368151.
15. Cody EM, Reagan AJ, Mitchell L, Dodds PS, Danforth CM. Climate change sentiment on Twitter: an unsolicited public opinion poll. PLoS One. 2015;10(8):e0136092. https://doi.org/10.1371/journal.
16. Cox R. Environmental communication and the public sphere. Sage; 2013.
17. Cox R, Depoe S. Emergence and growth of the field of environmental communication. In: The Routledge handbook of environment and communication. Routledge; 2022. p. 13–32.
18. Dash RK, Dash AK. The new media and the promotion of ecological entrepreneurship. Media Watch. 2019;10(2):419–31. https://doi.org/10.15655/mw/2019/v10i2/49634.
19. Aisyah Dwityas N, Mulyana A, Hesti S, Briandana R, Mungi KP. Digital marketing communication strategies: the case of Indonesian news 'portals'. VIII(3):307–16. https://doi.org/10.35808/ijeba/517.

Glossary

Behavioural change refers to the process of influencing individuals or organisations to adopt more sustainable and environmentally responsible.

Chants are rhythmic, repetitive vocalisations used in rituals, ceremonies, or religious practices. They cultivate a sense of unity, focus, or connection to a higher power.

Conservation is the protection and preservation of natural resources, ecosystems, and species to preserve ecological balance and biodiversity.

Environmental communication is the field dedicated to disseminating information on the environment, sustainability, and conservation initiatives. It aims to elevate awareness, promote behavioural change, and stimulate dialogue around environmental issues.

Folktales are traditional narratives passed down through generations within a certain culture or civilisation. These tales often include mythology, folklore, or supernatural elements and are intended to convey moral lessons or entertain audiences.

Legends are semi-historical accounts down through generations that elucidate certain events, places, or heroes within a civilisation or community.

Myths are conventional tales that explain the origins of the world, natural phenomena, and cultural practices. They often include deities, heroes, or supernatural beings and serve as a method to understand the mysterious.

Oral history is the recording and preservation of historical information via verbal accounts and personal recollections. It provides insights into the lived experiences of individuals and communities that may not be recorded in written records.

Rituals are a series of actions performed in a certain sequence, sometimes laden with symbolic meaning. They denote significant occurrences or transformations within a community or religion.

© The Author(s), under exclusive license to Springer Nature Switzerland AG 2025

G. Mutinta, *An African Perspective on Environmental Communication*,
https://doi.org/10.1007/978-3-032-05052-6

Storytelling refers to the oral or written process of communicating and passing tales throughout generations. It is a tool used by several cultures to convey information, provide amusement, and deliver moral instruction.

Sustainability refers to meeting present needs without compromising the ability of future generations to meet their own needs.

Index

A
The agenda-setting theory, 184
Air pollution, 40

B
Bidirectional dialogue, 12
Billboard, 61
Books, 64

C
Climate change, 37
Conceptual metrics, 186

D
Dance, 109
The diffusion of innovation theory, 183
D'Sa, Desmond, 88

E
Elders and wisdom, 106
Energy accessibility challenges, 45
Entertainment education, 20

F
Folk tale imaginary image, 98

G
Globalisation, 6

H
The health belief
model, 179

I
Industrialisation, 5
Interactive communication, 18

L
Land degradation, 32

M
Maathai, Wangari, 85

N
Nakate, Vanessa, 87
Natural disaster, 8
Newspapers, 58

P
Plastic pollution, 45

© The Author(s), under exclusive license to Springer Nature Switzerland AG 2025
G. Mutinta, *An African Perspective on Environmental Communication*,
https://doi.org/10.1007/978-3-032-05052-6

R
Radio image, 54
Rituals, 107

S
Saro-Wiwa,
    Ken, 84
Singing, 101
Stages of change
    model, 180
Storytelling, 95

T
Tribal meetings, 105

U
Urbanisation, 33

W
Water pollution, 36
Water scarcity, 34
Wildlife trafficking, 43

FSC
www.fsc.org
MIX
Papier | Fördert
gute Waldnutzung
FSC® C083411

Kolibri 360 GmbH
Ferdinand-Jühlke-Straße 7
99095 Erfurt, Deutschland
produktsicherheit@kolibri360.de